D0204311

*Individualizing reading
in the elementary school*

Individualizing reading
in the elementary school

DOROTHY RAYMOND

Parker Publishing Company, Inc. West Nyack, New York

© 1973, by

Parker Publishing Company, Inc.

West Nyack, N.Y.

*All rights reserved. No part of this
book may be reproduced in any form
or by any means, without permission
in writing from the publisher.*

LB
1573
.R274

Library of Congress Cataloging in Publication Data

Raymond, Dorothy MacLean.
   Individualizing reading in the elementary school.

   Includes bibliographical references.
   1. Reading (Elementary). 2. Individualized
reading instruction. I. Title.
LB1573.R274   372.4'147   72-10052
ISBN 0-13-457044-8

Printed in the United States of America

The value of an individualized reading program . . . .

Teachers who have observed children in an individualized reading program are impressed by the mutual respect evident among pupils and teachers, and the friendly helpfulness as pupils work together. The variety of creative activities reflecting wide interests and genuine enthusiasm for reading have caused many teachers to say, "'I want to teach that way. How do I do it?"

The purpose of this book is to show specifically how you can set up an individualized reading program. It can be started in any classroom at any time of the year. The very nature of the program would suggest that no two would ever be alike. Too much structure would lead to a dull routine that might stifle creativity. Without some plan, though, learning might not take place. For this reason, and completely aware of the inherent dangers, instructions will be given in a straightforward, almost prescriptive manner. It is hoped that they will give you the support needed to begin a new style of teaching.

Our reading program is individualized in that
—children read at different levels of interest and difficulty in
   material of their own choice;
—it is structured according to individual needs;
—children decide on and pursue a great variety of follow-up
   activities;

—many methods and techniques are employed according to the unique learning modes of the children.

One of the most important features is the individual conference with the teacher. Each child is enabled to establish a personal relationship with an interested adult while sharing reading experiences. Over and over children mention this as the most valuable and rewarding part of the program.

If you have been teaching with basal readers—three reading groups and quiet seat work—you may find it difficult to visualize yourself in a classroom where everyone is engaged in a different learning activity. It is hoped that specific instructions and illustrations will give you confidence to abandon the basals and teach reading in a more exciting and stimulating manner. To capture the spirit of the program, look through this book, paying particular attention to the illustrations of children's work. They originated in classrooms where teachers are ever anxious to provide creative learning experiences, and where projects are initiated by the children themselves. The variety is an indication of the interesting possibilities. The first five chapters contain ideas for various classroom procedures. You will note that the programs at different grade levels are more alike than different. The most pronounced difference is in the approach to those who can read and to those who cannot. Because of the many questions that have been asked by primary teachers more detailed instructions have been given for these grades. Actually any of the plans could be used at any grade level. The wording of certain instructions and the types of records kept would depend on the maturity of the pupils. These differences will be apparent in the examples given. Children need not be grouped by grade or age but because most schools are organized by grade level, programs are described for kindergarten through grade eight and for children with learning problems.

In addition to the specific grade level programs, more details and suggestions are offered in further chapters organized by topics. You will find instructions for determinig reading levels and individual needs, suggestions for teaching reading skills and ideas for creative follow-up activities. Since the pupil-teacher interview is vital to the success of the program, sample interviews are included along with techniques that have worked in our classrooms. A practical viewpoint is offered on record keeping by teachers and pupils and suggestions are made for evaluating pupil progress. Generally each topic is considered sequentially in order to focus on the developmental nature of learning.

There are many excellent articles and books available describing reading programs and offering specific suggestions on how to teach reading. Indeed, many of the procedures used in our school system have come directly or indirectly from sources cited in the bibliography and possibly from others difficult to identify. The basic ideas grew out of the author's attempt to apply clinical practices to the classroom but apply them before children fail. Briefly this means to find out how children learn and then provide experiences that will make it possible for them to do so. All teachers want to do this and it is most frustrating to know that some children need more help or time than the teacher can provide. In an individualized program, all children receive the amount and kind of help they need when they need it. Every child works at his "instructional" level and is aware of his own instructional needs. Teaching becomes diagnostic, progress is continuous, and no one experiences failure.

The program described in this book began in September of 1964, in a small city in central Maine where over half the population is of French descent. For many pupils, French is the first language heard and only an occasional child enters kindergarten able to read. In general, the median scores of our pupils would be average in most areas of national achievement tests. At first because of lack of classroom space and then because of budgetary difficulties, the pupil-teacher ratio tends to be high. The two original classes, a fifth and sixth, averaged 36 pupils. The most outstanding classes have continued to have enrollments of 30 to 35 pupils.

We have always made an effort to keep the community informed about school events and polices. Every new child is interviewed and individually tested and his room assignment and academic strengths and weaknesses are discussed with his parents.

Teacher participation in the individualized reading program, which now extends from kindergarten through grade eight, is with the exception of kindergarten completely voluntary. About 80 percent of the teachers and over 2000 pupils are involved. The names of those who have contributed specifically to this book appear in footnotes throughout. Grateful acknowledgment is made to them and also to administrators whose understanding and cooperative attitude eased the way for change—particularly Sarah K. Scott, Teacher and Principal 1946–67, whose support grew out of her deep concern for children; Buford L. Grant, Supt. of Schools, and Albert S. Hall, Ass't Supt., whose policies have encouraged teachers to be creative and innovative while striving for quality education for all.

DOROTHY RAYMOND

Contents

Individualizing reading
in the elementary school

1

Organizing yourself for individualized reading

Understanding the whole program

Trying to describe an individualized reading program is somewhat like the explanation of the Caucus-race in *Alice's Adventures in Wonderland:* "The best way to explain it is to do it."[1] In some ways an individualized program *is* like a Caucus-race for everyone begins at his own level and proceeds at his own rate. Thus, everyone feels successful.

> First it marked out a race-course, in a sort of circle ("the exact shape doesn't matter," it said), and then all the party were placed along the course, here and there. There was no "One, two, three, and away!" but they began running when they liked, and left off when they liked, so that it was not easy to know when the race was over.[2]

A dynamic flexible approach to teaching is easier to experience than to describe. Children tell it this way:[3]

> An individualized reading program is when you can read any book you want to and not read out loud in groups from any one

---

[1] Carroll, Lewis. *Alice's Adventures in Wonderland and Through the Looking Glass.* N.Y.: The New America Library of World Literature, Inc., 1960.

---

[2] Ibid, p. 33.

---

[3] Fourth, fifth and sixth grade pupils of Madeline Kenniston, Ellen Lancaster, Ida Nelson, Hazel Salsbury, Bette Townsend and Janet Weymouth.

reading book. It's when you pick out a book you like, read it, and make a small report on it such as the title, author, source and your comments about the book. You also put down any word you don't know and look it up in a dictionary. You can read at your own speed and can choose what you like and you have conferences with your teacher.

Individualized reading is exactly what it's called—you read books individually, alone! Nobody is made to read one book. It is very enjoyable. In individualized reading the teacher has fun, too.

The individualized reading program is really good (Photo 1-1). For instance, some people can read hard books and some can't. For the people who can read hard books, there are hard books, and for the people who can't read hard books, there are some that are easy. There is a good choice of books so a person could go to the book shelves and find a book that he would like to read. You are able to read at your own speed.

Photo 1-1.

Our reading begins as soon as we get inside. We begin by reading individually. We have one person who is the librarian. That way we keep track of books. We also have reading skills and the teacher helps us. There are many projects. My project is making maps with a group to show where stories took place. We have creative writing too. We do things like plays, stories, limericks. Some people are writing children's stories.

There're two hang-ups about individualized reading. No. 1—You don't get to have conferences with the teacher after every book. Like you could read a book, go have a conference with the teacher and she'd ask you questions about the book and you would talk about it. That would be perfect. No. 2—Records I hate because I love to read and I don't want to stop to do records.

Individualized reading is fun, easy and to put it into one word I *Great!*

Individualized reading gives you a chance to read what you want to read. You get to read any kind of book like fantasy, mystery, non-fiction, fiction, biographies, autobiographies, poetry. In individualized reading you read what you want. This is good because the teachers don't push you.

Individualized reading is that we may read what we want to read instead of being assigned something. What I dislike is that we have to stop reading at a definite time. But there is nothing we can do about that. (Is there?).

Individualized reading is every person reads his or her own book individually. You can choose your own book. You don't have to read a book. You can read a magazine. The teacher can see how much you're reading by checking your reading records. She can also see if you need help (Photo 1-2).

Photo 1-2.

In an individualized reading program, children select their own reading material according to their needs and interests as well as abilities. Books are shared with classmates in many varied and creative ways. Teachers work with individuals and groups in developing common reading skills. Instruction is systematic for each child and he is encouraged to capitalize on his unique mode of learning. An important feature of the program is the individual conference in which each child discusses the books he reads with his teacher. Children also enjoy reading to their classmates and pursuing interests in types of literature, authors, and subject areas. Because of the great enthusiasm for reading that develops it is not unusual for an average child to read and react to over a hundred books in a school year. A permanent interest in reading results.

A visit to a classroom would reveal many different activities. These vary with the age and maturity of the pupils and with the leadership and creativeness of the teacher. Teachers develop individual teaching styles but you will find certain types of activities common to all programs. During the reading period most children will be reading silently while the teacher is holding conferences. Some children will be keeping their records, working to improve reading skills, reporting on books read, practicing for a puppet show, making a mural, painting a picture or sharing learning in some other way. Some primary teachers devote one reading period to just reading and quiet-type activities in order to be certain that all children spend the major part of their time reading. All teachers will spend some time working with small groups. These may have been chosen to work on some common need—a comprehension difficulty or word recognition problem; or to discuss stories, story characters, interesting words, or in some other way help children make personal use of their reading. Most teachers begin and end each reading period with a whole group activity. This is the time when the class decides on organizational procedures and when children tell about their reading. At other times, the whole class might work together on such topics as noting the format of their books, using an index or table of contents, studying how to use dictionaries, etc.

An individualized reading program does not mean thirty groups instead of three. It does not mean that each child "goes through" identical learning procedures at different speeds. It does not mean that children always work alone. There is actually more group interaction than in the traditional classroom. It does not mean that pupils do only what they want to do. It does mean a different way of teaching. Above all, an individualized program enables children to feel successful and retain their self respect (Photo 1-3).

Photo 1-3.

Accepting a different role for yourself

An analysis of the twenty-seven studies of the Cooperative Research Program in First-Grade Reading Instruction[4] led to the conclusion that there is no one best method or set of materials to teach reading. Since none proved superior the responsibility rests with the teacher. You will be aware of this responsibility more in an individualized reading program than in any other. Be guided by the needs of your pupils, not by some artificially imposed standard. It will be necessary to know more about teaching reading because the range of reading abilities will actually be recognized. Your concern will not be getting through a book on time or covering a group of skills. You will know exactly which skills are necessary to help each child become a better reader and these will be the ones you teach. The emphasis will be on applying skills not verbalizing them as though the teaching of skills was an end in itself. If a child can pronounce most any word, you would not busy him with worksheets designed to check on his knowledge of vowel rules, for example. On the other hand, if he has difficulty with multisyllable words, you will teach something about syllables even if he happens to be in first grade. Thus, what you teach will be different, but more important, you and your pupils will know why it is being taught.

Another difference in your role will become apparent immediately. You will not be the focus of everyone's attention and you

[4] Bond, G. L. and Dykstra, R. *Coordinating Center for First Grade Reading Instruction Programs.* (Final Report—HEW Project No. X001) Minneapolis: Univ. of Minnesota, 1967.

will not know what everyone is doing at all times. Children will still look to you for guidance but they will become less and less dependent on you. Since you will spend most of your time working with individuals and small variable groups you will move about the classroom more. Some children will even be unaware of your presence. Your relationships with children will be more natural and they will treat you as a respected friend. They will be more anxious to ask for help knowing that your interest in their difficulties is for the purpose of helping rather than grading them.

Having children sit at desks in rows, although not impossible, does not belong with this style of teaching. Some teachers prefer tables and some have even put their own large desks out in order to make better use of the space. Enough space does become a problem. There will be times when small groups of pupils will be in the hall, auditorium, library or other available rooms. You will not be able to watch over all your pupils at all times and you will share them more with other teachers.

Because of the lack of ready-made materials and the range of reading levels in your classroom you will spend a lot of time choosing and reading books and locating or devising exercises and worksheets for your pupils. This will be particularly time-consuming your first year, but it is not an insurmountable task. It can be an interesting and most rewarding experience when you team up with other teachers to make these resource files (Photo 1-4[5]).

Photo 1-4. Resource file in 4th grade room.

[5] Larry Littlefield.

Thus you will no longer be the dominant person in the classroom asking the same questions and leading everyone through the same learning procedures year after year, but you will enjoy your new role of helping and supporting active inquiring pupils.

A fifth grade teacher[6] in her sixth year of an individualized program describes the change in her role:

> The basic change in my role teaching in an individualized reading program is that I feel I am more of a helper than a teacher. Instead of standing over a whole group and wondering if everyone is being reached, I know what each child needs and he works on that skill by himself.
>
> I have found I have to be more creative in my teaching as it is individualized and must meet the needs and levels of all the students. Teaching is therefore more interesting and stimulating to me.
>
> I know each child better because of this program. In the conferences, I not only am able to give more help but feel that the child and I have a more friendly relationship.

A third grade teacher[7] in her second year says:

> One of the most amazing aspects of an individualized reading program is how the role of Teacher changes from that of director—even dictator!—to that of advisor, observer and confidant.
>
> Too often in the basal approach to reading, Teacher chooses, Teacher motivates, and Teacher attempts to be ringmaster in a three-ring circus. The premise is that if a teacher is conscientious about covering all the material and channels it in the right direction, readers will somehow absorb just the right amount of information and skills. Because in individualized reading children select their own reading materials, keep their own records, and choose their own follow-up activities, they are self-motivated rather than teacher directed. What, then, is Teacher's place in the new scheme of things?

---

[6] Janet Weymouth.

---

[7] Patricia Myers.

The focal point of the individualized reading program is the pupil-teacher conference. Here, in a special one-to-one relationship, Teacher becomes a *friend;* getting to know the child, learning likes and dislikes, abilities and disabilities, and nurturing what will hopefully become a warm relationship. As a *statistician,* Teacher collects and evaluates all available background information about the child such as achievement scores, reading levels, interests and aptitudes. It is in the conference that Teacher becomes a *diagnostician,* pinpointing problem areas and giving individual attention to skills when needed. Here, too, Teacher may become *advisor* while assisting the child in the selection of activities and projects to be done independently, or with others for sharing purposes.

In the area of skills there is the special obligation of knowing the skills that should be taught, having the needed materials readily available and being able to teach them individually or in small groups at the right time. Hence, Teacher must be an *organizer,* able to have flexible groups and able to reorganize groups and materials when needed.

Although children in individualized reading do much of their own record keeping and help with the organization of materials in the room, Teacher must also be *secretary,* keeping track of teacher and pupil records, worksheets, books, magazines, projects, skills files and art materials needed for the preparation of sharing projects.

Teacher also has the never-ending job of *procurer* of books and materials since a wide variety of reading matter on many subjects and many levels is essential to the program.

The teacher of individualized reading plays many roles. Basically, however, children are doing the "telling" for a change, and Teacher is "listening!" As a good *listener* Teacher has a perfect opportunity to know, understand, and encourage children as individuals. By being able to communicate in a friendly and understanding way, and by showing genuine interest in individual needs, the teacher of individualized reading is better able to safeguard the child's self-image as an important and contributing member of society.

A sixth grade teacher[8] put it briefly: "The teacher's role changes in an individualized program from talking and directing to listening and suggesting."

---

[8] Ellen Lancaster.

Getting the classroom ready

There is no reason why an individualized reading program cannot be carried on in any ordinary classroom but certain arrangements may make it easier. These are best worked out with the class. The following suggestions may prove helpful.

Arrangement of room

The traditional idea that every child has to have a specific desk in a specific place in a classroom is unfortunate. It is impractical to be moving furniture around every time groups are formed and unless the room contains work areas and extra chairs and tables you will feel cramped for space. You might arrange the desks so that some are in groups of four, some in pairs and some single. Pupils will keep their belongings in a certain desk but will understand that they may not always be working at that desk. If your classroom has desks nailed to the floor, remove some or all and replace them with tables. Pupils without desks can store their belongings in sectional shelves or in plastic containers kept on shelves. Try to arrange the furniture so that one corner of the room becomes a library where children can browse and enjoy books. In primary rooms particularly, this area should have a carpet.

Acquiring books

Since most teachers are not familiar with the reading levels and interests of their pupils before school begins it will be necessary to start with a wide range of reading materials. This problem is alleviated somewhat in schools containing a central library but each classroom must have its own books. Hard cover books are beautifully illustrated and well worth their cost but it is probably unrealistic to suggest purchasing them when getting an individualized reading program under way. We buy mostly paperbacks. They are not only inexpensive but liked by children. Be sure that the money usually spent on basal texts and workbooks is used to purchase classroom books. The average cost of workbooks alone for a second-grader is \$1.65. This would come to \$49.50 for 30 pupils —more than the cost of 100 paperbacks. If you also considered the price of tests and additional materials that accompany a basal series the average per pupil cost goes up to \$3.00. For a class of 30 that would be \$90 or 200 paperbacks. If yours were a new class, you would need to budget \$5 per pupil in addition for the two basal readers. This would add \$150 more! When there is a limited amount of money available purchase books for the poorer readers as they will depend more on you for their books than will the others.

The public library is a good source of books. Make arrangements to borrow fifty to a hundred for a month. Where possible make the selection yourself. You might exchange boxes with another teacher and cut down your trips to the library. Get the parents interested in helping; they can go during school hours. Children should be encouraged to get books from the library on their own, join a book club and bring books from home.

It is sometimes difficult to get enough reading material for the poorer readers. Select stories from literary readers, social studies and science books and old basals. Cut them out and bind them in attractive folders. These are particularly valuable for older pupils who need a great many experiences with material at approximately the same level of difficulty. You could add suggestions for follow-up activities.

Remember to add newspapers and magazines to your collection.

Storing books

The books should be kept away from traffic areas. No special order is needed in the primary grades but some teachers like to use the same procedure as in the school library. In any event, no grade designations or labels such as "easy" should be used. Learning how to choose a book is one of the aims of an individualized reading program.

It is better not to put out all the books at once. Some might be introduced along with a topic related to science or social studies and you will want to keep certain ones for special holiday reading. Pupils look forward to new reading material and it helps to make those exciting days before vacation profitable. An example of this will be found in Chapter 8.

Preparing yourself for conferences and other activities

It is not necessary to have read every book your pupils will read. Primary teachers probably should but it would be an impossible task for upper grade teachers. If the stories don't interest you, consider it as a course in speed reading. You will gain more than just an acquaintance with children's stories.

There are three things you might want to do with some of the books. If you intend to use "pupil checkers" you will need to make out word lists and questions. The usual procedure is to have pupils ask each other the routine factual type questions. A "pupil checker" can be anyone who has read a book and has had a conference about it with a teacher so everyone can be a checker. It encourages children to help each other and gives them an opportunity to discuss books. It also frees the teacher to concentrate on vocabulary and inferential type questions and to help with reading problems during the conference. In the primary grades, instead of questions, a child might read all or part of a book to a friend or to a small group before discussing it with the teacher. Questions pupils are to ask each other should bring out an understanding of the story, and it should be possible for them to check the answers by rereading. Actually our pupils read to one another, discuss books in small groups and work together to report on books, but no classroom uses pupil checkers as a regular part of the program.

The following might be used by a pupil-checker in second grade for WHAT DO YOU DO, DEAR?[9]

Say these words and phrases:

library	rubbers	treasure chest
ranch	downtown	around the fire
quietly	tightwire	floating away
mouth	unexpected	white fur coat
bookmark	lovely	walk the plank

Which words above are compound words? (bookmark, downtown, tightwire)

Questions (to be read orally by the pupil-checker)
    1. What do you do when the pirate's cook says, "Luncheon is now being served?" (Wash your hands before eating.)

---

[9] Joslin, Sesyle and Sendak, Maurice. *What Do You Do, Dear?* New York: Young Readers Press, Inc., 1969.

2. When were you to find a bookmark to save your place? (When the Sheriff of Nottingham appears to take you to jail.)
3. Why do you have to go out in the rain? (To save a Princess.)
4. What should you do when a lady drops something? (Pick it up and return it to her.)
5. What do you do when a large and hungry dragon comes to a birthday party? (Thank the hostess before leaving.)

Now take turns reading your favorite parts to each other.

Secondly, you might also want to make out questions or outline points to bring up during a conference. These could be filed on cards. As you use the suggestions you can delete or add to make future conferences more interesting. Such a system would be a great help to a substitute teacher. The opening question should be an invitation for the child to tell you what the story meant to him. It follows that one set of questions could not be used for every pupil. Perhaps it's best to have a two or three sentence summary of the story to remind you of the important points followed by statements or questions for possible discussion. Is Johnny Tremaine[10] the type of person you would want for a friend? Why or why not? How did your opinion of Johnny change after certain events which you or the pupil might mention? These events could be listed on your card. Most books can be read and discussed at various levels. A fifth grader might enjoy the narrative without getting the implications of the American Revolution. Your discussion will reflect this.

The third thing you might want to do is to provide suggestions for follow-up activities. The ideas should be unusual or unique to the story and children should never feel they have to follow them. A natural follow-up for *The Snowy Day*[11] would be: Read the book to your friends and then all of you make a mural or write a story on how you would have fun in the snow. Another might be: If we have snow, read the story to the class and suggest they go out and try doing some of the things Peter did.

Books containing obsolete information can help a child realize that everything in print isn't absolutely true. *Goodby, Tree!*[12] could lead to a reexamination of the use of DDT. Sometimes try pro-

---

10 Forbes, Esther. *Johnny Tremaine.* N.Y.: Dell Publishing Co., 1969.

11 Keats, Ezra Jack. *The Snowy Day.* N.Y.: Scholastic Book Services, 1966.

12 Woyke, Christine. *Goodby, Tree!* N.Y.: The Macmillan Company, 1967.

viding the items necessary to carry out the project: For *The Hundred Dresses*[13] have an envelope of brightly colored paper and a few paper dolls. You won't even need to make any suggestions. Your ideas may not all be accepted but they will insure variety.

Specific sources of children's books and classroom materials

When purchasing books we prefer individual titles of our own choice rather than classroom sets and we seldom buy more than one copy of a book for any one classroom. The teachers choose their own books which are considered instructional materials and therefore are in addition to those purchased for the school library.

The following annotated list of sources is offered to help you get started.[14] (Children's paperbacks begin at 35¢ and average 50¢ so that 100 books would cost about $50.)

Paperback and Softbound Books

Ace Books, 1120 Avenue of the Americas, New York, N.Y.   10036
Particularly science fiction for upper grades.
American Book Co., 300 Pike Street, Cincinnati, Ohio   45202
Whitman, Thomas. *See and Say! Picture Books.* These twenty picture story books are designed to help young children develop oral language facility. We use them to encourage pupils of all ages to write stories to accompany the pictures.
Jacobson, Willard J. et al. Units for *Thinking Ahead in Science,* Grades 1–6. Short science units available in paperback with no grade designations.
Manolakes, George (ed.) *Reading Round Table Series.* 54 titles in paperback PP-6 may be purchased individually or in boxed classroom sets. Books contain comprehension checks and word study exercises.
Bantam Books, Inc., 666 Fifth Ave., New York, N.Y.   10019
Books on sports and cars particularly good for upper grade boys.
Pathfinder Editions—"the best in fiction and non-fiction in a wide variety of subject areas."

---

[13] Estes, Eleanor. *The Hundred Dresses.* N.Y.: Harcourt, Brace & World, 1944.

---

[14] For more comprehensive lists cosult texts on the teaching of reading; for up-to-date news of books see advertisements in professional journals like *The Reading Teacher* and *Journal of Reading* (published by International Reading Association) and the Book Review section of *The New York Times.*

Bomar Publishing Co. 622 Rodier Drive, Glendale, California   91201
Tiny books for primary children.

Dell Publishing Co., Inc. 750 Third Ave., New York, N.Y.   10017
Yearling Books—outstanding children's books by distinguished authors for grades 2–8.

Laurel-Leaf Library—for upper elementary.

All available as individual titles or Classroom Library Units (Unit H, for example, are books related to American History).

Educational Reading Service, Inc. East 64 Midland Ave., Paramus, New Jersey   07652
Individual titles and classroom libraries for K–8. Each set contains 100 different titles, circulation cards, and book ends. Special curriculum related collections including books accompanied by records for beginning readers. Sponsors book fairs and classroom book clubs.

Garrard Publishing Co. Champaign, Illinois   61820
Humphrey, James and Moore, Virginia. *Read and Play Books.* Games and stunts children read and demonstrate. *Harper & Row* P.O. Box 98, Elmsford, New York   10523
American Heritage illustrated paperbacks for elementary and junior high.

Science Unitexts—68 booklets on single, specific topics in biological and natural sciences that can be used in any grade. Reading levels primary to junior high.

Real People Series—43 biographies of 36 pages each for grades 5–8.

Hertzberg-New Method, Inc. East Vandalia Road, Jacksonville, Illinois   62650
Literature Exploration Laboratory. Thirty-five books covering the major literary works influencing contemporary American thinking, teacher's guide and student study cards. Comprehension questions at four levels of difficulty. Aimed at grades 10–12.

Learning Materials, Inc. 600 Madison Ave., N.Y.   10022
The Literature Sampler—Samples selections from 120 books in areas of courage and daring, sports, people, animals, fun and mystery. Comprehension questions and pupil record books. Junior and Senior editions. We use the junior edition in grade 5.

Macmillan Co. 866 Third Ave., New York, N.Y.   10022
Reading Time Books. Soft covered books designed to be used after specified chapters of their basal series. Each book is one complete story.

The Spectrum of Books. Set A & B with 30 in each set for grades 4 up. Plasticized covers.

Noble & Noble Publishers, Inc. 750 Third Ave., New York, N.Y.   10017
Crossroads Books of short stories and poems—junior high up.

Falcon Books for junior and senior high.

Discoveries in Literature—Paperback study units high school.

Scholastic Book Services, 904 Sylvan Ave., Englewood Cliffs, New Jersey   07632

Readers' Choice—Catalog of over 900 paperbacks K–HS, curriculum units, study skills books, map skills project books, book-record combinations and classroom libraries for grades 2–6. Book clubs: See-Saw, Lucky, Arrow, TAB, Campus.

Classroom Units for individualized reading for grades 3–6 including 100 titles, book case, teaching guide, conference cards for teacher, activity cards and reading logs for pupils.

Reluctant Reader Library for grades 6–7 includes 50 titles and teacher's guide that gives synopsis and three questions with answers for each book.

Scott, Foresman and Company, 99 Bauer Drive, Oakland, New Jersey   07436

Eight Starter Books and Eight City Starter Books for kindergarten or first grade.

What's Happening and *Something Else.* Paperback anthologies of contemporary selections for eighth graders with 5th and 6th grade reading ability.

Simon & Schuster, Inc., 630 Fifth Ave., New York, N.Y.   10020

Papertexts grades 5–12 with discussion questions, vocabulary review, suggested composition activities.

Young Readers Press, Katonah, New York   10536

Individual titles or sets for grades 2–6.

In addition, the *American Education Publications,* Education Center, Columbus, Ohio 43216 publishes the Weekly Reader Paperback Book Clubs. The teacher's bulletin contains activities for motivation, discussion and follow-up.

Hardcover Books

The American Publishers Corp., Educational Reading Service and Follett Library Book Company handle books from virtually all book publishers. Their catalogues are organized by grade (K–12), interest, and subject areas. They are the best comprehensive source of hardcover books for children.

American Publishers Corp., 1024 Washington Blvd., Chicago, Illinois   60607

Educational Reading Service, East 64 Midland Ave., Paramus, New Jersey 07652

Follett Library Book Co., 1018 W. Washington Blvd., Chicago, Illinois 60607

The following companies offer boxed classroom sets:

Follett Library Book Co., 1018 W. Washington Blvd., Chicago, Illinois
   Classroom libraries emphasizing literature, social studies or science.
Harper & Row, P.O. Box 98, Elmsford, New York   10523
   School Readiness Treasure Chest—sets of picture cards and 36 books
   Torchlighter Library 44 books K–3
   Torchbearer Library I 37 books (3–5)
   Torchbearer Library II 34 books (4–6)
   The American Adventure Series—not boxed but indispensable
Holt, Rinehart and Winston, 383 Madison Ave., New York, N.Y. 10017
   The Owl Program—KIN/DER, Little, Young, and Wise
   Holt's Impact Series (grades 7–9)
McGraw-Hill Book Co., 330 West 42nd St., New York, N.Y.   10036
   Literature, science and social studies—8 boxes (K–9)
Random House, 201 East 50th St., New York, N.Y.   10022
   Carousel Books and Reading Pacesetters—a supplementary individualized program including reading card sets for each book—vocabulary, comprehension, activities, skills.
Scott, Foresman and Co., 99 Bauer Drive, Oakland, New Jersey   07436
   Talking Storybook Boxes (K)
   Invitations to Story Time (K)
   Invitations to Personal Reading (1–6)

Three companies that carry many books of high interest (K–12) but low reading levels are:

Benefic Press, 10300 W. Roosevelt Rd., Westchester, Illinois   60153
   Butternut Bill, Button Family, Moonbeam, Tom Logan, Dan Frontier, Sailor Jack, Cowboy Sam, Space Age, Mystery Adventure and *Sports Mystery* (Most series begin at PP level)
Field Educational Publications, Inc., 396 Springfield Ave., Berkeley Heights, N.Y.   07922
   Kaleidoscope, Checkered Flag, Time Machine, Jim Forest, Morgan Bay Mysteries, Wildlife Adventure, Deep-Sea Adventure, Reading-Motivated and Americans All.

Children's Press, 1224 West Van Buren St., Chicago, Illinois 60607
Up to 71 titles—average about 30—in each set as follows: Easy-Reading Picture Story, Reading Laboratory, I Want to Be, At the Books, About Books, True Books, Indians of America, Frontiers of America, Cornerstones of Freedom, Pictorial Encyclopedia of America, Science Discovery, Enchantment of America State and Regional, Tizz Books, Young People's Story of Our Heritage, The You Books, True Story Biographies, Adventures in Nature and Science, Let's Travel, Exploring the World, People of Destiny, Fun to Read Classics and The Animal World in Color.

You will need a variety of dictionaries including a college level one for yourself.

American Book Co., 300 Pike St., Cincinnati, Ohio 45202
Webster's New—Elementary, Practical, Students and Collegiate. Also Third New International.
Children's Press, 1224 West Van Buren St., Chicago, Illinois 60607
Picture Book Dictionary and Dictionary of Basic Words
Follett Educational Corp., 1010 West Washington Blvd., Chicago, Illinois 60607
Follett Beginning to Read Picture Dictionary
Golden Press Inc., 830 Third Ave., New York, N.Y. 10022
Golden Dictionary, Golden Picture Dictionary, Illustrated Golden Dictionary, My First Dictionary, My First Golden Dictionary, Picture Dictionary for Children, Richard Scarry's Best Word Book Ever, Richard Scarry's Storybook Dictionary.
Harper & Row, P.O. Box 98, Elmsford, New York 10523
Words I like to Read and Write and Words to Read, Write and Spell
Holt, Rinehart & Winston, 383 Madison Ave., New York, N.Y. 10017
Very First Words for Writing and Spelling, Pixie Dictionary, Word Wonder Dictionary, The Holt Basic Dictionary of American English, The Holt Intermediate Dictionary of American English, The Winston Dictionary for Schools.
The Macmillan Co., 866 Third Ave., New York, N.Y. 10022
My Self-Help Dictionary, My Word Clue Dictionary, Webster's New World Dictionary.
Random House, 201 East 50th St., New York, N.Y. 10022
Picture Dictionary
Scott, Foresman and Co., 99 Bauer Drive, Oakland, New Jersey 07436
My Little Pictionary, My Second Pictionary, Thorndike-Barnhart Beginning. Junior, Advanced and High School Dictionaries, In Other Words I and II, A beginning and junior Thesaurus

2

A basic plan of classroom organization for individualized reading

The purpose of this chapter is to present a view of classrooms at work without going into the methodology of teaching reading. We begin in kindergarten and first grade with a language-experience approach. In these rooms, nearly all reading is from group and individual charts. As children begin to read books the classroom procedure gradually changes. Because pupils are engaged in many different activities, the organizational pattern is not always evident and teachers and observers generally ask about it. This topic, therefore, will be considered before specific teaching techniques. There is no *one* way to organize a class but knowing about successful plans might help you decide on a workable plan suited to your class.

What the children do

First and foremost those who can, read (Photo 2–1). In the primary grades, reading is scheduled for one hour in the morning and one hour in the afternoon. In many traditional classrooms only the teacher is engaged in reading activities for the whole two hours, but in an individualized program everyone has meaningful reading experiences. At any one moment, nearly every child in the room may be doing something different but there usually is a basic pattern: choose a book, record the name of the book and author, read the book and while reading keep track of any difficulties encountered such as word recognition problems or inability to understand, record the number of pages read each day, have a conference with a teacher, react to the book in some personal way, and work on specific reading skills if necessary.

A child may confer with another child or with a teacher or librarian when choosing a book but record keeping and reading are individual activities. At the primary level very few children are aware of comprehension difficulties but they readily seek help from fellow pupils or a teacher when they come to an unknown word. The follow-up on a book may be an individual presentation particularly if it is written but more often two or more will join for an oral report, dramatization or puppet show. Children will read to one another or to a small group and in rooms where pupil checkers are used they will hear pupils answer questions over a story and perhaps pronounce and give the meanings of selected words and phrases. Work on specific reading skills for most pupils will be both individual and in groups. Every child will have the privilege of a conference alone with his teacher. There are many opportunities for group interaction and because children do have unique contributions listening becomes worthwhile.

Photo 2-1.

Children who have not learned to read will follow a different procedure. They may compose stories using their own file of words, picture dictionaries, word lists and charts displayed in the room

5

(Photo 2–2 and 2–3). They may ask another pupil or a teacher for a word. They will read each other's stories and listen to stories read by other pupils. Many varied experiences will be necessary to help them gain verbal facility and supply them with ideas for writing group charts. Proportionally more time will be spent in oral talk and discussion about the experience than in writing the chart.

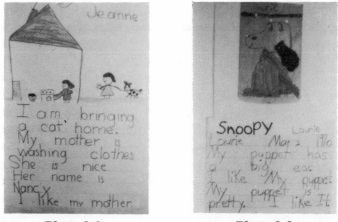

Photo 2-2.   **Photo 2-3.**

One such plan is to provide an experience, discuss the experience, develop a group chart, have pupils make their own experience or reading charts, and then bring the whole group together to share the individual charts. These may then be stapled together to make a book which can be read and enjoyed later. When the group is made up of readers and non-readers they may all share the same experience; the better readers then read about the topic while the non-readers make experience stories and the beginning readers either read or write a story. The whole group eventually comes together for further discussion. In all these plans, pupils are surrounded by many rich verbal experiences.

What the teacher does

The teacher will take every available opportunity to teach reading. The unique feature of this type of program is that each child can get the specific help he needs at the time he feels the need. Most of the teacher's time in class is used for individual conferences and the rest in small group instruction. There will probably be group discussions and a daily sharing period. In some classrooms, the pupils begin reading activities as soon as they enter the room. After the teacher has taken care of such preliminary activities as attendance reports, she may begin conferences, bring a small group together for instructional purposes or invite the whole class to participate in a language activity. This might be rather broad, like sharing humorous events from stories or telling how an incident in a story is similar to a personal experience. Or it might be more specific like preparing for a lesson on syllabication by asking each child to tell the longest word he can find in his book. The teacher would continue with conferences and end the class period by having individual pupils or groups report on their reading.

The conference requires thoughtful preparation. You will begin by reviewing your records of the last conference and checking the pupil's records to see how much he has accomplished and difficulties he has had. Then you will *listen* as the child tells about his reading. He may read a selection orally and you will ask questions and have a discussion to bring out word meaning or comprehension and word analysis skills. The conference will end with definite plans for future reading activities. You will write up each conference and keep a brief record of your daily activities just as the pupils do. This latter should be reviewed and evaluated frequently to make sure that you are making the best use of your time. Because you will know each child's instructional level and specific needs you will be able to help him at strategic times throughout the day. This is not to imply that reading instruction be left to chance. There must be a definite instructional plan for each child.

Establishing classroom procedures

The sooner pupils assume responsibility for directing their own learning activities the better. You must be free to work with individuals and small groups without interruption. Begin by listing the things that have to be done by the class as a whole and by individual pupils. You could do this in ten minutes and in a few more minutes could assign duties to certain pupils, but this is not the way to help pupils to be responsible. Have everyone contribute to the list and decide whether each job is an individual or group responsibility. A list for a second grade might be:

INDIVIDUAL	GROUP
Keep a record of daily reading.	Check file box for order.
File folder in box in alphabetical order.	Keep classroom library in order.
Check bulletin board for notice of conference.	Collect and check books to be returned to library.
Have folder ready for conference.	See that paper, paint, crayons, scissors, etc. are stored neatly.

The jobs carrying group responsibility will be rotated at stated intervals. Have two pupils on each job; one to be in charge and the other to help. Teach the procedure to the first two, but from then on replace only one. The helper takes charge and the new pupil becomes the helper. In that way, they teach each other and soon everyone in the class knows every job.

Most individual responsibilities require following directions. If children are seated in small groups while you give the directions *once,* they can help each other. You could move about the room to clear up any difficulties and answer questions a group has. Try a variation of the following to introduce the topic in an interesting manner.

Each pupil will need a piece of paper 8″ square and a pencil.
Each group will need a transparency and a grease pencil.
Give oral or written directions:
1. Fold your paper so that two opposite corners come together making a triangle.
2. Place the triangle on the table in front of you so that the fold is toward you.
3. Now fold the triangle in half placing the left side of the triangle over the right side.
4. You now have a smaller triangle. Fold that in half.
5. Your triangle is still smaller. Fold that in half.
6. Now open up your paper.
7. With a pencil, trace over all the creases.
8. Compare with others in your group. Are they all the same?

9. Choose one you agree is correct and trace it on a transparency.
10. Put the transparency on the overhead projector to compare with those of the other groups.

Project the correct design on an overhead projector and have the transparencies from each group placed over it as a check. Evaluate. Why were the groups successful—or why not? If they were not, let them try again after the evaluation. Follow this immediately with something related to the reading program such as instructions for keeping a daily record of reading. Give each child a duplicated record form, give directions for recording the title of the book and then walk about the room to observe how well they are able to do this. If anyone has a question, suggest he try to get help from someone in the group before asking you. The results are sometimes unexpected. This is the way one second-grade child recorded Anita Brenner's *A Hero by Mistake*.[1]

NAME OF BOOK	AUTHOR	PAGES READ
A Hero	Mistake	

Children who have experienced small variable group work from kindergarten on find it very easy and quite natural to work together. Others may have difficulty cooperating at first. Some of the techniques used in group dynamics are helpful if the activity is followed by discussion to bring out the feelings of members of the groups. One such activity is illustrated in Diagram 2–1.[2] The problem is to have five people cooperate in constructing identical squares.

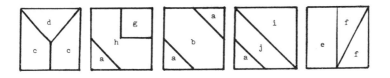

Diagram 2-1

[1] Brenner, Anita. *A Hero by Mistake*. Katonah, N.Y.: Young Readers Press, Inc., 1966.

[2] Cartwright, Dorwin and Zander, Alvin. *Group Dynamics Research and Theory*. Elmsford, N.Y.: Row, Peterson & Co., 1960, 679–680.

1. Divide the class into groups of five.
2. Give each group five envelopes containing shapes—
   Envelope A—Shapes i, h, e
   Envelope B—Shapes a, a, a, c
   Envelope C—Shapes a, j
   Envelope D—Shapes d, f
   Envelope E—Shapes g, b, f, c
3. Explain that each envelope contains pieces of puzzles that can be put together to make five identical squares. Each person in the group will receive one envelope which he will open and empty in front of him. Then all will start to form squares. There are three rules that must be followed.
   a) No person may speak to another.
   b) No person may take a piece of a square from another or in any way indicate that he wants it.
   c) Anyone may give one or more of his pieces to another.
4. When a group has five perfect squares all the same size they stand.

Observe the pupils as they work to make the squares. Then have them react to a variety of questions to help them express their own feelings and learn how others felt. You might want to begin by having each group tell how they went about the job. Some groups might be unsuccessful because there are many ways of making wrong squares in that such squares prevent the formation of the other four. Once a person has made a square he is reluctant to take it apart and he may sit quite oblivious to the fact that the others need his pieces. Ask questions like, "How did you feel when you saw a piece you needed and the person didn't give it to you? How did you feel when you saw the first group stand? How did you feel when you thought the other members of your group couldn't solve the problem?" The final question might be, "What do we learn from games like this?"

There are many times when you may have small group discussions. Difficulties arise when teachers assume that children know how to do this. Techniques for carrying a fruitful discussion must be taught continually. Some interesting topics can be chosen just for this purpose. At other times, after the objectives of the discussion have been taken care of, the techniques themselves can be evaluated. Did everyone understand the topic; did everyone say something about it; did the others listen; did the group agree on a summary; how could your group have had a better discussion?

Having flexible groups without confusion

When asked how she liked substituting in an individualized reading program a teacher said that at first she had never seen so much confusion in her life, but then she realized that she was the one who was confused because the pupils were all engaged in constructive activities. Actually the noise level and movement and interaction of pupils varies greatly from day to day and from teacher to teacher. The program is individualized for teachers as well as for pupils.

If you prefer a highly organized classroom, you will be more comfortable if everyone including yourself works by a routine schedule. You probably will have a special time for silent reading when everyone except a pupil having a conference stays seated and reads. You will have another time when everyone works on reading skills. Then most pupils will be at their seats while you work with small groups. You will have another scheduled time for projects when children are allowed to talk quietly to one another. There will be a specific time each day or week when pupils share their reading with each other or with the whole class (Photo 2–4).

Photo 2-4.

If you thrive on confusion, and some people do, you will have a variety of activities going on at once. Some children will be reading while others work on projects. Having discovered during a conference that one child needs help on some specific skill he may be doing on a worksheet. You might go on from there to work with a small group or you might walk about the room and have a word of encouragement for everyone and end by calling the whole class together for a sharing session.

In an individualized reading program, class organization is important only as a reflection of the teaching atmosphere. This is very difficult to describe. A common observation made by visitors is that there is no tension in the rooms. Children feel free to make errors and reveal their problems. Particularly in the primary grades they will invite visitors to listen to a story, offer to give a puppet show or just come over and quietly ask for help with a word. Perhaps this atmosphere comes about because of the conferences where pupils and teachers share their feelings. Perhaps it's due to the personal involvement each child feels toward his learning. Much of it is a reflection of the teacher's feelings. Children are very adept in this area of reading. They read facial expressions and tone of voice. They also read feelings. When the teacher is bored the children are bored. When you are pleased with them they are pleased with themselves. Some people are naturally better teachers than others because they start off with warmer personalities and greater empathy for children. An individualized reading program will not automatically make everyone a better teacher. The point is that flexibility enables teachers and children to respond to each others' feelings. This relieves the pressure and tense situations are avoided.

You will find that group work will evolve naturally when groups are formed in response to a definite need. If you have six children who continually mispronounce medial vowels, call them together and tell them you think they all have the same problem with words. Get them to identify the problem. Then you can all get down to work. You might take them for the first ten minutes every day as long as they need you. As each child solves his problem he leaves the group and eventually the whole group dissolves and a new one takes its place.

It is not necessary to have just individual and small group instruction. Dictionary skills and so-called comprehension skills might better be introduced to the whole group. Use material that all can read. After teaching the skill, have them apply it to their own books.

For example, in teaching how to state the main idea, write or adapt paragraphs that even the poorest reader in the class can understand. Afterwards have each pupil choose a paragraph from his own book and be prepared to tell you or write the main idea.

In general, groups that you will work with directly are more effective and cause less disruption if you plan them for the beginning of a period.

Photo 2-5.

Encouraging children to help each other

When pupils are no longer competing for grades or the teacher's attention they willingly help each other (Photo 2-5). There are many opportunities for this in an individualized reading program. Pupils will ask each other words when reading and writing. When they read one another's stories the author is there to help with words and the reader may note errors in spelling and punctuation. Groups can be given pictures or a few items and asked to discuss various story possibilities. Each person then writes his own story and reads it to the group. In a second grade class, two girls[3] wrote a play and made six copies of it all on their own. (What meaningful practice in penmanship!) Then they chose characters and helped them learn to read their parts. They presented it to their own class and later to a third grade class.

Whenever pupils work together on projects they learn from one another. We encourage our older pupils to write stories and poems for the younger children. Some of these are placed in the library and others are presented to classrooms. This poem was one of many

---

[3] Elizabeth Brody and Beth Reuman.

stories and poems written by fourth graders for kindergarten children. The request came about because there was so little available on the topic.

SUPERMARKETS[4]

Supermarkets as you can see
Are made for people like you and me.
It's there we go to get some food
Whenever we're hungry or in the mood.
They have different things to eat
Like potatoes, beans and lots of meat.

Another group[5] wrote and illustrated a poetry book about animals for first graders who spent many hours reading and rereading it.

I KNOW A LITTLE SEAL[6]

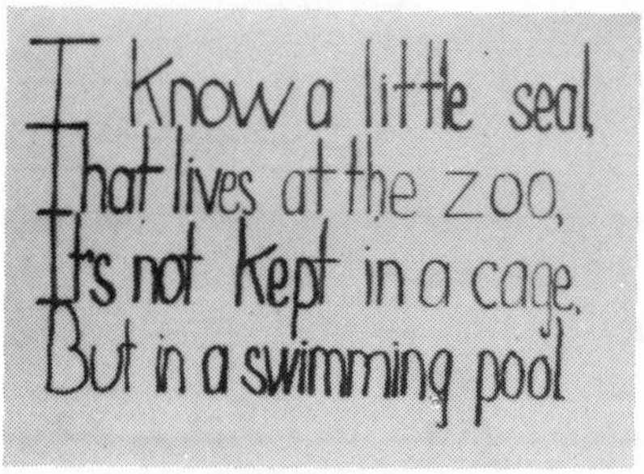

I know a little seal,
That lives at the zoo,
Its not kept in a cage,
But in a swimming pool

Photo 2-6.

---

[4] Paul Carey.

[5] 4th grade—Madeline Kenniston, teacher.

[6] Janice Tash.

THE TASMANIAN WOLF[7]

Once there was a Tasmanian wolf,
Who always howled a rof, rof, rof
And when he made a howl,
He always seemed to growl.
Now people seem to say,
"He's only a Bluff!"

HORSES[8]

Horses are ever so quick.
They always know more than one trick.
Horses get angry, horses have fun.
Horses can jump and kick and run.

THE DEER[9]

Deer are so full of fear.
I wonder why they're so dear?
Deer are ever so funny.
I wonder why they're so happy?

Photo 2-7.

---

[7] Michael Osborn.

[8] Larry Lavway.

[9] Yann-Ling Pan.

Photo 2-8.

It is important that every child be given a chance to help another. Actually this is not difficult for even the non-readers write their own unique stories and so use words that others may not know (Photo 2-7). Older pupils who for one reason or another have not learned to read very well not only bring enjoyment to young children but teach themselves to read when they write children's books (Photo 2-8).

3

*Specific ways to begin an individualized
reading program*

The range of reading ability in a classroom is much greater than most people realize. Even in kindergarten where children are not expected to read there will be some who can read stories written for second and third graders. In second grade, there will be some children who can hardly read a word but the best readers will have reached fourth and fifth level. Special planning is required to meet the needs of all but the first problem is to get started. Five plans are offered. Your choice will depend partly on personal preference and partly on the general reading ability of your class. Illustrations will be given from various grade levels. Since the plans are quite flexible they could be used for a change of pace at any grade level throughout the year.

Starting with a common experience

When pupils and teacher share a common experience they have something to talk about. If the experience is worthwhile it may be worth recording. This may be in the form of a group experience chart, individual experience charts, individual stories or a combination of all three (Photos 3-1 and 3-2).

Photo 3-1.

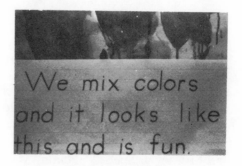

Photo 3-2.

Photo 3-3.

Some children may read while others are coming to the understanding that anything that is said can be written and read. An experience could be a science experiment, making something,

(Photo 3-3), discussing a problem or doing something like taking a trip. It is suggested that they be organized around broad under-standings or be the kinds of experiences which would lead to con-ceptual growth. The general procedure is to experience something, talk about it, decide what might be written, write it, and then read it. Procedures for a class trip will be outlined.

1. Assuming the purpose for the trip has been defined and understood, divide the class into heterogeneous groups of three or five to discuss procedures to be followed on the trip. This gives everyone a chance to express an opinion. Circulate among the groups to help them keep to the topic when necessary but do not enter into the actual discussion.

2. Bring the whole group together to exchange views. Suggest that a chart be made listing procedures or rules for the trip. Get the title of the chart from the group. Lead the discussion but the actual wording will reflect the maturity of the group. As you write this and succeeding charts it is important to have the language objectives in mind.[1] By appropriate comments and questions to the group (How does *bus* begin? We just used *we* in the last sentence; how do you spell *we?*) you will not only teach but assess each child's progress.

3. You may want the class to follow a similar procedure to list the objectives of the trip or list questions they expect to have answered.

4. After the trip, give adequate time to discuss it. Help the children organize and express their thoughts orally. Then whether they write or dictate to you they can concentrate on writing and reading.

5. A variety of group and individual experience charts will now provide many reading experiences. Children who can read should be provided with appropriate books.

Photos 3-4, 3-5 and 3-6 illustrate a visit to a supermarket by a kindergarten class.[2] Their specific purpose was to note the different jobs performed. This was related to the broader concept that people are dependent upon one another.

---

[1] See Chapter 4.

---

[2] Ruth Wood.

A second grade class[3] had fun sliding downhill on pieces of cardboard. They wrote an illustrated book, *Cardboard Fun.* Some captions were:

They are sliding on the cardboard. Two cardboards are going down the hill. They are having fun.

Robert took a Jump. Paul and I Crashed. Elias is on the top.

They are sliding down the slippery hill.

We are sliding down the hill. Weeeeeeeeeeeeeee!.

Sliding over the hills.
It is fun. You go fast.

Bill is chasing Paul.

I am sliding on the hill in a cardboard box. It is fun.

Paul is in front. I'm in back. Paul went over a bump. (Photo 3-7[4].)

The store lady picked up a
live Maine lobster for us.

Photo 3-4.

An approach through literature

Sometimes we begin a lesson with poetry or a story. The following suggestions may be adapted to either.

1. Read a story to the whole class about a summer activity such as *A Day of Summer.*[5] The children will enjoy telling about things they have done. Then suggest they each draw a picture to

---

[3] Lucille Winters.

---

[4] Joni Amalfitano.

---

[5] Mitchell, Lucy Sprague. *A Day of Summer.* N.Y.: Alfred A. Knopf, Inc., 1960.

illustrate people having fun in the summertime. Tell them the pictures will be put together to make a book. While the children are drawing sit down with the various groups to talk about the pictures. Accept all pictures and stories without criticism or comparison. As children complete their pictures sit apart from all groups and have each child come to you individually to tell you more about the picture and decide what is to be written about it. This may be a word, phrase, or sentence. Write what the child says and have him read it to you. For some, this will be an experience chart. Others will actually read. After a child has done this, direct him to books you have chosen and have him take one to his seat to look at or read. To avoid confusion when everyone will be getting books for the first time, allow children to share books with the people around them but not to keep going back and forth to the library area. After the pictures have been completed, walk around and enjoy the comments of the children in order to give everyone a chance to look at a book. Conclude by taking one of the shorter stories and reading it to them. One first grade class[6] had these ideas:

I made a tree house.

I'm going swimming in the pool.

I threw the ball over there.

The moon is out.

Me and my dog are doing a show.

I'm looking at the birds with binoculars. (Photo 3-8[7])

I'm hanging up the clothes.

I'm lying on the grass looking at the clouds.

Me and Randy went to the swimming pool.

2. Before the next class session, fasten the pictures together using a blank sheet of paper for the cover. Tell the children you are going to read a book to them—the one they wrote. Read it with the same enthusiasm you have for library books. After a discussion, ask what would be a good title for the book. Write the suggestions on a chart. Without being critical, keep reminding them that the title usually tells what the book is about. (Keep in

---

[6] Janet Mitchell.

---

[7] Patrick Theriault.

mind, but don't verbalize to the class, that you are building an understanding of the main idea.) The class above suggested:

The Summer Day

The Summer Night

The Summer Morning

Fun in Summer

The Summer Book

Fun Day Book

Summer Fun

Have the children decide on a title (they chose *The Summer Book*) and write it on the cover. Place the book on the library table and suggest they look at it or read it and if they need help they get it from the person who drew the pictures. Children will continue looking at or reading their books. Tell them you will come around and ask what their book is about. Accept all answers. You might want to ask a question to help children clarify their thoughts. For example, a child might say her book is about a little girl. Ask what the little girl did or ask some other appropriate question. When the class begins to get restless conclude by making a general remark about the variety of books. You could mention categories and have children raise their hands if their book fits the category. Tell them that tomorrow they can draw a picture of one of the characters from a book and put it on the bulletin board. Then read them a story.

Photo 3-5.

Photo 3-6.

Photo 3-7.

Photo 3-8.

3. If you have beginning readers, continue with the language-experience approach. If not, tell the class that you want to find out

how many words they know so you are going to have them come to you one at a time to say some words. You are going to call them in alphabetical order. While you do this they may read or draw a picture of a character from a book. Suggest that someone might choose the book the class wrote. Begin checking words.[8] Your immediate objective is to make a quick check for instructional level as far as words are concerned. You will complete this informal reading inventory during conferences after you have the class organized to work on their own. It may take a few sessions to check everyone particularly if you have to stop from time to time to help the class. Allow time to have book characters cut out, identified and placed on the bulletin board. Make labels for each and have them placed by the pictures. Actually most children will make their own labels copying the name from the book. (Photo 3-9)

4. As you continue checking words children may look at or read books, add to the bulletin board or illustrate the book they are reading. Some might be encouraged to write something about the characters they chose.

Whenever you introduce a new activity take time to be sure everyone understands it. Your objective is to help pupils become self-directed.

5. After all the words have been checked, you have a *tentative* instructional level based only on word recognition. The purpose is to help you determine the immediate word recognition needs of each child. This knowledge along with the lists of "new" words encountered in their reading will form the basis for specific instruction. You are now ready to begin your program and continue with the informal inventory.

Photo 3-9.

---

[8] See Chapter 9 for word lists and the procedure.

Using pictures

Pictures can be used at any grade level. How they are used will be determined in part by the reading and writing ability of your pupils. In presenting the suggestions, non-readers will be considered first. In general, children in a language-experience approach gradually turn to books as they find they can read them. Surrounding the children with many interesting easy-to-read books will offer encouragement.

1. Using Pictures for Group Experience Charts
   a) Show a picture to the group. This might be a slide of one of their own activities. Encourage the children to tell stories about the picture. After considerable oral exchange of ideas, suggest that a story be written. Make it into a group experience chart.[9]
   b) Divide the class into heterogeneous groups of five. Take each group separately to discuss a picture and write a story. This may take a few days. After all the stories have been written, bring the class together and have each group "read" their story orally. Have the class compare stories and help them conclude that different stories can be told about one picture.
2. Using Pictures for Individual Experience Charts
   a) Have children paint or draw pictures; tell about them and dictate a word, phrase, sentence or paragraph about them (Photo 3-10[10]).
   b) Display a variety of pictures. Let pupils choose one to write about. They would then read each other's stories, getting help if needed from the author. A seventh grade class[11] wrote from silly stickers such as, I Dare You to Mess My Hair, Fragile I'm Wearing Glasses, Member Stupid Club, Hot Lips, You Dog, Cereal Grows Hair, I Love Me. Their imagination is unlimited. This one carried the sticker HELP![12]

---

[9] See Chapter 4.

[10] John DeWitt.

[11] Gary Lisherness.

[12] Dorcas Benner.

Little Johnny Ant was playing in his miniature ant hole, when all of a sudden a huge spider popped out at him.

Little Johnny just sat there staring, because he had never come across a spider before. But as the spider raised one of his awful legs, Johnny let out an awful scream. "HELP! HELP!" "There's a monster looking at me!"

"Aw, come on," said a squeaky little voice belonging to the spider. "I was only waving at you. Would you like to play with me? I'll let you see my bug collection," the spider said hopefully.

"Well, I guess I can play. But only if you promise not to squish me."

The spider nodded, and the two walked leg in leg (or whatever spiders and ants have) to the spider's web.

c) Use picture story sequences such as those published by the American Book Company or Scott, Foresman and Company. Have children at any grade level supply the words. The following, written by children of various ages, was inspired by *Next Time*.[13] Note how a beginning second-grader[14] used a black crayon to emphasize some words.

Photo 3-10.

---

[13] Whitman, Thomas. *Next Time* (See and Say! Picture Books). Cincinnati, Ohio: American Book Co., 1969.

---

[14] Lisa Stewart.

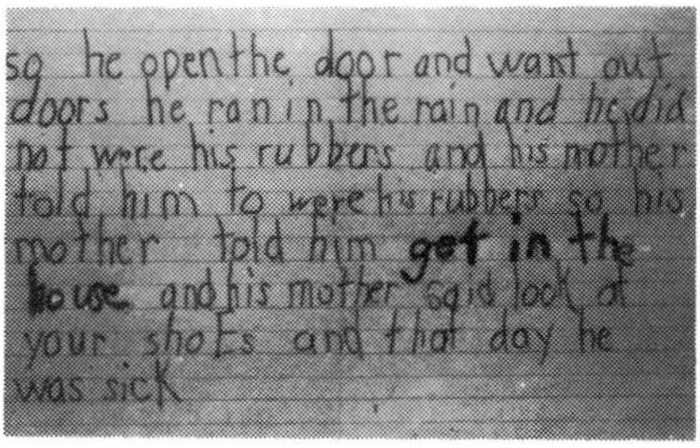
so he open the door and want out
doors he ran in the rain and he did
not were his rubbers and his mother
told him to were his rubbers so his
mother told him get in the
house and his mother said look at
your shoEs and that day he
was sick

Photo 3-11.

One day a little boy saw another little boy and it was Raining out and so he ran down to get his raincoat and hat and was going out and his mother came and said put your rubbers on. The boys face was mad so he hided them out back of the trash so he open the door and want out (Photo 3-11).

Different approaches by fifth and sixth graders.

It was a rainy day. Paul felt so bored. What could he do on a rainy day? Nothing, absolutely nothing. Oh, he felt terrible. Then he caught sight of someone walking in the rain. He decided that would be fun. Away he sped to put on his raincoat. He wanted to get away quickly so he wouldn't have to put on his rubbers. Away he went straight for the door. Oh no, his mother had caught him. She told him to put his rubbers on. She was mad because he had tried to avoid her. He felt mean and helpless. He was going to pay his mother back for being so mean. Instead of putting his rubbers on he hid them behind a paper basket.[15]

Tommy looked out the window and thought Oh boy I'm going out.

He then ran to the closet, opened it and took out his raincoat, put it and his hat on and started to the door.

"Tommy" his mother said, "You forgot these" holding out his rubbers.

---

[15] Lydia Cauz.

Tommy thought as he looked at them, "I don't want to wear these." Quickly looking around he hid them behind the rubbish can.[16]

One day last summer, Kenneth was sitting by a window.

"Rain, rain, rain," said Ken, "that's all it's been doing this week. "Hey wait a minute. I could go outside and jump in the puddles. Better get my raincoat."

"Just a minute young man," said his mother, "If you go outside you have to put your rubbers on!"

"Ugh! Rubbers," said Ken, "maybe I can hide my rubbers. Let's see, I guess they'd fit behind the wastebasket.[17]

It was a cold, rainy day and Danny wasn't having any fun inside at all. So he decided to go out and play in the rain. He ran to the closet and got his rain coat and rainhat on, and then went to the door. Just as he opened it,

"Danny!" the voice boomed "put on your rubbers!"

"Aw Mom, do I have to?" Danny asked back.

Now Danny was very sly and he sure didn't want to wear rubbers, so, he put the rubbers in back of the garbage pail.[18]

One day Tom was looking out the window at the rain. He decided to go out to splash in the puddles so he went to put on his rain things. He put them on and skipped to the door. But Mother got there ahead of him. She held out a pair of small brown rubbers, and told him sternly to put them on. Tom looked angrily at the rubbers, then gave a sly look at the retreating figure of his mother. Quietly, he hid them in a good hiding place, and tiptoed out the door to freedom—without rubbers.[19]

> I looked out the window,
> And what did I see?
> Rain, rain, rain,
> Oh, golly gee!
>
> I ran to the closet,
> And opened the door.
> I put on my coat,
> And scampered across the floor.

---

[16] Cathy Spencer.

---

[17] Brian Letourneau.

---

[18] Susan Mavrinac.

---

[19] Peter Phair.

I put on my rain-hat,
And walked to the door.
I started to open it,
But didn't do any more.

Mom had my rubbers,
Clamped together with a pin,
And if I didn't put them on,
I'd have to stay in.

I looked at my rubbers,
I scowled and I said,
"I won't wear my rubbers,
I'd rather go to bed!"

Then I had an idea,
Hide my rubbers, why sure!
Then before Mom can see me,
I'll run out the door!

But Mom saw my rubbers,
And boy was she mad!
She called me in,
And she said I was bad.

But never again,
No, never I said!
Will I ever be bad,
And get sick in bed![20]

And from a seventh grade girl:

Mother cannot understand
Why I'm not acting gay,
It's raining cats and dogs outside,
And I want to go out and play.
I know what I will do,
I'll go out anyway.
"A little water does no harm"
Is what they always say.
My coat is in the closet,
I'll get it right away.
Boy, I'm going to have some fun,
Out in the rain today.

---

[20] Becky Lane.

My hat is on, I'm all bundled up.
Am I ever going to have fun.
I cannot wait to get outside,
Maybe I should run.
"Oh, wait a minute David,
You cannot go out yet."
"You haven't got your rubbers on,
You'll get your feet all wet."
Boy, does this ever make me mad.
Why do I have to wear these?
If I don't wear them out,
Mother will never be pleased.
This is a good place to hide them,
Mother will never look here.
Now it's time to get outside,
This rain won't last all year.
Yippee! This rain is really fun.
I've never done this before.
I hope Mother doesn't find my rubbers,
Behind the wastebasket on the floor.
Oh, Oh, Look who's waiting for me,
Mother's at the front door.
"Next time, David, you'll listen to me.
You won't play in the rain anymore."
Mother was right you know.
Next time I'll do as I'm told.
So don't go out and get all wet,
Or you'll end up in bed with a cold.[21]

Beginning with children's own reading

The easiest way to begin when all children in a room can read is to have each one choose a book and read. Book selection should be left entirely to the individual. Some beginning readers find this such a difficult task that it is a good idea to have children discuss how they choose a book. Suggestions could be written as a group experience chart to be left on display. More suggestions will be made if the children talk first in small groups. Whether it be that way or by whole class discussion expect and accept such responses as: Read

[21] Danette Proulx.

the title to see if it's about something you want to read about, look at the pictures, look at the words to see if they're easy, see if it has more words than you want to read. Establish the idea that they check for interest and difficulty. The first is rarely a problem; the second may be for some. Suggest that one way to decide is to open to the middle of the book and try to read a page. Beginning readers having difficulty with three or more words should probably put the book back and try another. Advanced readers, able to draw on more word recognition skills, may be able to use five words as the criterion. This can be worked out in individual conferences. The choice should be left to the child though. Teachers sometimes find it difficult to do this. When older pupils first go from a basal program into individualized reading it is a common occurrence to have the good readers choose very easy books and the poor readers, books much too difficult. The good readers seem to welcome the chance to enjoy reading while the poor readers strive to improve their self-image. This situation does not last very long and it is best to treat it rather casually.

When a second grade class[22] began individualized reading we told them to take a book they thought they would like, open to any page and read orally to one of us to be sure the book would be easy enough. We found it unnecessary to make suggestions. Some children read a sentence or two and decided their book was too easy while others stumbled over words and concluded theirs was too hard.

A sixth grade class presented more problems. One girl who had a very low reading level claimed that no book interested her. She was finally asked to choose one of two books, read it and then tell why she liked or did not like it. An individualized reading program always offers choices, but not the choice to be a non-reader.

During the first few days you might:
1. Help children choose books.
2. Invite children to read portions of their books to you.
3. Begin systematic check on instructional levels and needs of each child.

During class sessions pupils will spend more time reading than in other activities. Introduce only one new procedure at a time. In

---

[22] Evelyn McDonough.

the following suggested plan, the first three steps might take one reading period. The fourth and fifth steps might take a week.

A suggested plan

1. Tell the class they are going to have a different kind of reading program—each person will choose whatever he wants to read. Ask how to choose a book. After a few suggestions, divide the class into heterogeneous groups to discuss this problem and come up with a list of helps.

2. Bring the class together and make a chart of their suggestions.

3. Have each child choose a book. Give help where needed.

4. Invite each pupil to show you his book, tell why he chose it, and read a page or paragraph to you. Your objective is to assess but not criticize his ability to choose.

5. During the conference have each child make out an appropriate record sheet for the book. This may indicate name of book, author, and place to note new words.[23]

6. Children continue reading as you hold individual conferences.

7. Provide a folder where each pupil can keep his records. Print names on folders (last name first) and file alphabetically. Each child should be responsible for filing his own folder.

8. While children continue reading, begin a systematic check of each child's strengths and weaknesses. The best way to do this is with a reading inventory.[24] Since reading is taught both morning and afternoon in the primary grades you might give reading inventories only in the morning.

9. After all children have been checked, follow one of the basic plans or a variation better suited to you and your class.

Teaching reading through subject areas

What do children read? In most traditional reading classes, they read stories of daily life and an occasional folk or fairy tale. Little,

---

[23] See Chapter 11.

---

[24] See Chapter 9.

if anything, is learned from the content but the greatest criticism is that this type of reading is inadequate preparation for other areas of study. In an individualized program children learn to read all kinds of material. Sometimes they enjoy all one type such as plays, mysteries, biographies. At other times everyone might read different types of material all related to one topic or historical period. Generally though children should be completely free to choose their own reading material during the reading period. Particularly when using language-experience approach, the experience could be related to the subject matter fields.

How a kindergarten teacher[25] helped a class gain a better understanding of voting is an example from the social studies. Because Election Day followed shortly after Halloween, the teacher had pupils make and cast ballots to decide how their classroom pumpkin should look—sad, happy or scary.

Photo 3-12.

1. This class had gone on a trip to purchase a pumpkin and now it was displayed on a table.

2. Pictures of Halloween pumpkins were shown and discussed and the class concluded that most pumpkins looked happy, sad or scary.

3. Each child was invited to draw a picture showing how he thought the class pumpkin should look. The picture would be folded and placed in a ballot box.

---

[25] Ellen Reed.

4. After the balloting, the class gathered around the box in front of a chart. The teacher wrote the three words (happy, sad, scary) on the chart and made a tally as each picture was taken from the box and classified. When the class couldn't decide, the person who drew the picture provided the answer.

5. Then the group counted the "votes" to determine the winner.

6. They watched while the teacher carved the appropriate face (Photo 3-12).

4

Overcoming special difficulties
in an individualized reading program

Many teachers have accepted the idea of an individualized reading program in theory but feel they lack certain knowledge or skills to be successful. The two areas mentioned frequently are making and using experience charts and knowing what words and "reading skills" to teach. A common remark is, "The children love the program but I'm not sure I'm doing everything I should." This chapter will focus on the problems most often identified by both those who are in an individualized program and those who are thinking about starting one.

Developing group experience charts

There is no one best way to develop and write an experience chart. Find the ways that are most comfortable for you and meet the needs of your pupils. Everyone in the group can learn something but more will be learned when the teacher has planned objectives for individual children as well as for the group as a whole.

Making a chart should be considered an important learning activity. This means that all those involved must attend to the discussion, decide what is to be written and watch as it is written. If these conditions do not prevail, it is better not to make the chart. Children's attitudes toward reading will be favorable if chart making is held in high esteem.

There will be a variety of excuses for writing a chart. These are usually considered the purpose; the only real purpose from the teacher's standpoint is to teach reading. Using charts makes it possible to expose children to the complete cycle of language development in a natural and interesting way. They see that written words represent spoken words. When children read their charts there is no stilted word-by-word reading because they know how those words were spoken before they were written. This is an important point. The chart is the culmination of group talk. Because it is visible and tangible evidence that we taught something there is a tendency to have too many teacher imposed charts. Some children could benefit by more talk and less chart work. When they know how to express their thoughts orally they can concentrate on watching these ideas appear in print so that they can read them. This is not to imply that there is a "readiness" or waiting period. Every child should see something he says written every day.

There are always a few children entering kindergarten who for various reasons say very little or do not speak at all. They will need opportunities to "verbalize" their experiences. By having planned activities, children will have something in common to talk about. The teacher can encourage conversation by asking pertinent questions and making suggestions while the children are engaging in the activity. Later they can tell what they did and the teacher can help them choose some part to record. Early group charts are very often individual contributions. Children watch their own words, phrases or sentences written and at first may only remember their own (Photo 4-1[1]).

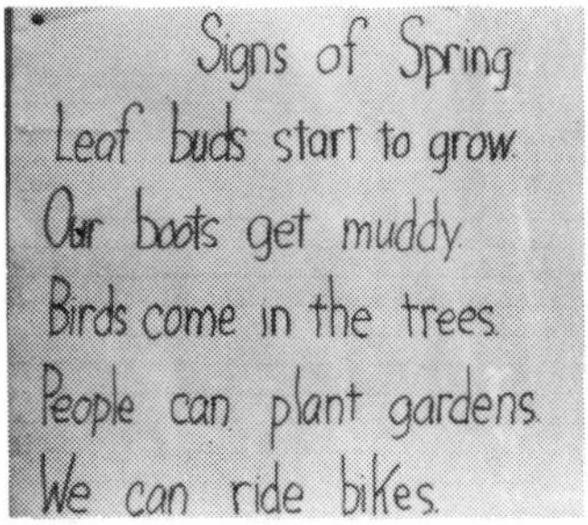

Photo 4-1.

---

[1] Vera Austin.

Photo 4-2.

Jamie   Karyn

Wendy   Beth Ann

Gary   Patty

Debbie   Bethany
clean up

Photo 4-3.

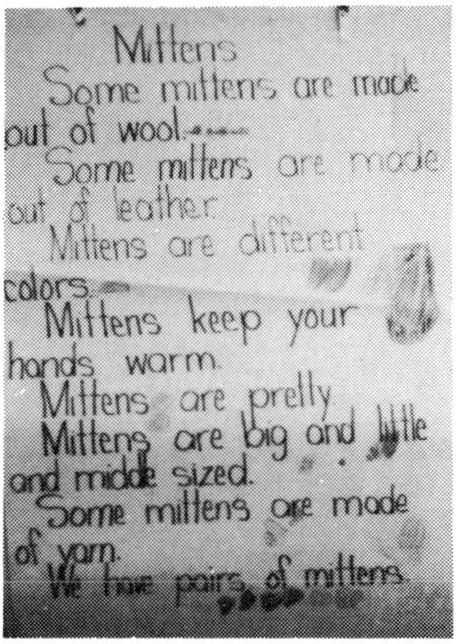

Mittens
Some mittens are made
out of wool.
Some mittens are made
out of leather.
Mittens are different
colors.
Mittens keep your
hands warm.
Mittens are pretty.
Mittens are big and little
and middle sized.
Some mittens are made
of yarn.
We have pairs of mittens.

Photo 4-4.

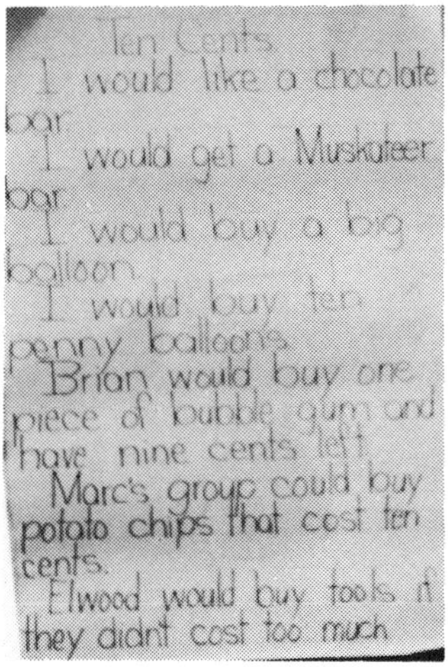

Ten Cents
I would like a chocolate
bar.
I would get a Musketeer
bar.
I would buy a big
balloon.
I would buy ten
penny balloons.
Brian would buy one
piece of bubble gum and
have nine cents left.
Marc's group could buy
potato chips that cost ten
cents.
Elwood would buy tools if
they didn't cost too much.

Photo 4-5.

Other early charts may be lists of words. Pictures can be added to help recall, and sometimes actual objects are used. See Photos 4-2[2] and 4-3.[3]

Later charts will reflect group cooperation and decisions. Keep them uncluttered. If a sentence takes more than one line, divide by phrases to help make reading meaningful (Photo 4-4[4]).

How does anyone learn anything? The teacher will have a general idea of what is to be written but the actual words will come from the children. To get the best responses without discouraging any child, have a number of children make suggestions. You might then restate some of them and ask if anyone can say it yet another way. By the time you actually write, each child will feel that he contributed. Just before writing you will say or get the group to say what you will write. Then all watch as you write. Another way to get all to participate is to have small groups discuss the topic first. In a first grade class,[5] the mathematics lesson was about money. The next day the teacher began the reading lesson by asking how many had ever had money they could spend any way they wished. She wrote the different amounts ranging from five cents to a dollar on the board and the various points that had been brought out in the mathematics lesson were reviewed. Then she divided the class into small heterogeneous groups to discuss briefly what could be bought with ten cents. The class was brought together again. Some children went off to write on their own while the others made a group chart (Photo 4-5).

As a chart is being written, the teacher will make comments to bring out an understanding of language which may be as general as, "Now I'm going to write what Billy said. Billy said . . . ," or as specific as, "The next word is *boat,* what letters shall I make first or Bobby can tell us how *boat* begins. Tell them how you knew. (*Boat* begins like *Bobby.*)" All the while the teacher has in mind an objective for the group which she will comment on, such as sentences begin with capital letters, and objectives for individuals

---

[2] Carole Daisey.

---

[3] Jayne Robinson.

---

[4] Beulah Churchill.

---

[5] Beulah Churchill.

based on the knowledge she knows they have. No child should be asked a question he obviously can't answer. The teacher's thinking may go like this:

> I'll begin by asking Tony to show us where to begin writing, I'll comment on the capital letter at the beginning of the sentence and get the whole group to tell me to end with a period. Brenda will be asked to spell certain words and many will spell in chorus words we have used often like *we*. Bobby and Maria will know most of the initial consonants. When I get to the last sentence if the group is not telling me to begin with a capital letter I'll ask them whether I should make a small or capital letter, etc.

Every child will be exposed to the language understandings while the teacher is constantly alert to what each child is learning. The following is how the first grade chart referred to above was started: Now let's find out what you decided would be a good way to spend ten cents. What would be a good name for our story? They chorused "Ten Cents" and the teacher wrote it with the comment that everyone seemed to agree. Someone immediately gave the first sentence. The teacher told them to note how she was spelling *would* and pointed out that *chocolate* was a long word. She asked if chocolate bars really cost ten cents (Yes!). Another child indicated he wouldn't get just any chocolate bar but would want a Musketeer bar. Others agreed and it was decided to put that down as the second sentence. The third sentence came as the answer to a direct question and the group was asked to spell *would* as it was written.

Usually each sentence is read just after it is written. Sometimes it is helpful to have the class read to decide a good ending. In any event, when a chart is completed it should be read immediately by all together. You may indicate the lines by a sweep of the hand or you may point to each word as you read along leading the group. What happens next will vary according to their needs. Two or three children may read the chart in unison or one may read alone. Pupils may be asked to read the sentence they like best, point to a word (either a specific word or any word just to see if they realize the difference between a word and a letter), find two words that look alike or begin with the same letters, find a word that begins like their own names, read all the words that begin with capital letters, tell how many words are in the first sentence, choose a sentence or word to illustrate, draw a picture that tells about the whole story, etc. The first grade class above continued this way: The teacher told them to look around the next time they went to a store to see what can be bought for ten cents and they would make a list. (They were reminded of this again before going home.) They then read the chart and different children read their favorite sentences. Some read the "number" words, the words with three syllables and the words with apostrophes. It was then suggested that each person draw a picture and write about what he would buy. They got their words from the chart, picture dictionaries, each other, and the teacher who moved about to give each child a new word. The teacher then called up small groups of children and had them reread with many purposes: a sentence that tells the size of something, a sentence with a number word in it, sentences that tell about something to eat, two sentences that tell about something to play with, etc. The words *by, buy* and *Elwood, wood,* and *would* were written on the board, compared, and discussed as were *did not, didn't* and *Brian* and *Marc's.*

In general when making a group chart, no attempt should be made to control vocabulary. The organization and tone will be a reflection of the suggestions made to the group and the types of questions asked. *Examples:*

> Let's list all the words you might want to use to write a Halloween story. (No order expected.)
>
> What did we use in our science experiment? Now let's list them in the order in which we used them.
>
> Our story is about our rabbit. Let's read what we have so far and see if each sentence tells something about the rabbit. . . . What else could we tell about the rabbit? So far we have just told what he looks like. Can you think of something interesting that he does?

Some teachers attempt to narrow the vocabulary to conform to certain lists thinking that children will learn to read faster. This is not necessary. Pupils will go from charts to literature and an author does not start out with a list of words when he writes a story. In the language experience approach, reading is more than memorizing words.

Most teachers find that lined newsprint lasts as long as they usually care to keep the charts. If it is to be recopied on better grade paper, you might do it with the help of a few children. They will enjoy reading and spelling the words for you. Sometimes a group chart serves to motivate individual stories and pictures which are assembled into a book (Photo 4-6). Then the group chart could be copied and included as the first page in the book. Other permanent charts include those placed around the room as a source of words for children's writing.

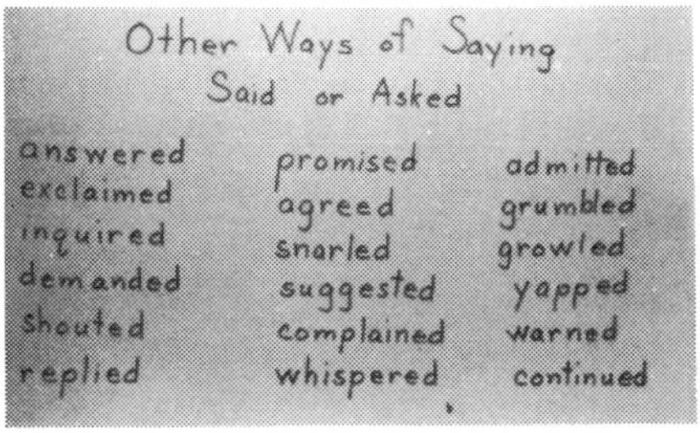

Other Ways of Saying		
Said or Asked		
answered	promised	admitted
exclaimed	agreed	grumbled
inquired	snarled	growled
demanded	suggested	yapped
shouted	complained	warned
replied	whispered	continued

Photo 4-6.

Using individual experience charts

Group work is valuable but there is no doubt that more learning comes from individual attention. When a child makes his own chart he not only has the individual attention of the teacher but the advantage of receiving just the help he needs. See Photo 4-7.[6]

Much the same method is used as with group charts. Be sure to have the child tell the whole story. Then ask what part he wants to see written. As you write he should be next to you so that he views the writing from the correct angle. Your comments will be a combination of teaching and checking and he will read it back immediately. As children develop an interest in writing the technique can be varied by having the child make a copy under yours or you write on another piece of paper and let him copy it onto the original. Some teachers make letters of dotted lines so the child can trace

---

[6] Kathleen Gray, teacher.

over them. Individual charts on the same subject can be stapled together to form a book. Each child can read his own page or the teacher could read it to the group. The book is then placed on the library table for all to enjoy. See kindergarten experience stories,[7] Photos 4-8 and 4-9.

Photo 4-7.

Photo 4-8.

Photo 4-9.

Kindergarten experience stories.

---

[7] Ruth Wood.

Choosing words

When children begin to take an interest in writing they are encouraged to write their own stories. They keep their words on cards. These serve as a record of words learned and as a source for correct spelling although there is no direct teaching of spelling. Other words come from picture dictionaries, charts placed around the room, classmates and teachers. Pupils read each other's stories. When the first graders wrote their stories about spending the only word that caused any difficulty was *taste*. The boy who used it asked six people at three tables how to spell it. Some hunted in dictionaries and one suggested it was like *paste* except for the *p,* but they decided it didn't look right. By the time the teacher arrived to write it on a card, he had stirred up considerable interest. Incidents like this are valuable learning experiences. He had drawn a big red apple and his story appears first below:

I like apples.
Apples taste good.
Apples cost 10¢.
Look at my good apple.

I am bringing a package.
It cost ten cents.

About a Dime

I would buy some bananas.
I would buy an apple.

I would buy some bubble gum.
I would buy some potato chips.

I am going to buy a popsicle.

I would buy soda pop.
I like soda pop.
I like all flavors of soda pop.

I would get some bubble gum.
Bubble gum can cost ten cents.

I am going to the store to buy a candy bar.
I like to eat candy bars.

I would buy a boat.

I am buying some chips.
I like to eat chips.

I would buy a ball.
I like it.

This is a ten cent candy bar.

We have found that writing is a successful technique with beginners of all ages. In first grade, we have a special time each day when all pupils are allowed to ask for one word. This word is printed on a card by the teacher. The child takes his word and uses it in a "story." At first, he may just draw a picture to illustrate it. The stories may include words from many other sources and he is never limited in his word choice. Words that he wants on cards, though, must be recognized the following day. If they have been forgotten, they are taken away because they would have no value. If the word is wanted again, another card will be made. The children accept this procedure because they understand that these are their own personal writing words. We do not keep reading words on cards because this would be impossible. Some children like their cards so much that they make many of their own but there is no competition to amass the most, probably because this is not mentioned. Actually the best readers will have the fewest cards because they soon learn to spell and become adept at using special charts and dictionaries (Photo 4-10). The following is an example of this.

Photo 4-10.

A girl in first grade was given a graded word recognition test[8] in early January. She scored 100% at 3rd grade level, 96% at 4th grade level and 88% at 5th grade level. Her word box contained over a hundred words. None were the usual words taught in first grade for she had no need for them. Here is a sample: chipmunk, gorilla, pudding, polite, primitive, nonsense, costumes, octopus, kayak, medical, Mississippi, travel, interesting, Kennedy and Daniel Boone. A boy whose reading level was beginning second grade level had among other words: ice, goal, skates, Boston, New York Rangers, Montreal, puck, tape, stick, hockey, teams, goalie, hat trick, overtime, National and Bruins.

The children keep their word cards in boxes. During the time set aside for writing they place their cards on tables and read them to each other or sort them in some way. Some first graders had noticed in reading their words that the long words took longer to say. This led to an interest in syllables. One youngster's words were arranged as follows (sample of a few):

bike	birthday	alphabet	mathematics	Indianapolis
paint	mermaid	grandmother	Rumpelstilskin	
horse	costume	kangaroo	Canadiens	
queen	angel	history	motorcycle	

After looking at their words the children begin to write. The teacher moves from group to group checking on the previous day's words, and from time to time all of the words, and gives each child the new word he requests. Many of these words are related to a group experience but some are purely personal. Some children will continue with word cards in second grade, but most will not need them. They do continue to write though. We also have them keep "diaries" which do not have to be seen by anyone, but the children do enjoy sharing them. They do this for about ten minutes a day usually in the morning or when school is getting under way after lunch. The following are samples:

> My frend and I went Outdoors and we climed my fence and jump from my fence into the snow. It was fun. We did it 5 times.

---

[8] See Chapter 9.

I had fun on my grandfather's Birthday. I did a head stand four the compane. Sunday we came home.

Today is a very messy day. I do not like today. When I got up my sister was crying. She was crying because it was too dark. I said, "It is not too dark." It was too dark for her because she was talking in her sleep.

Concerning "skills"

When children learn to talk they gradually acquire the grammatical construction intuitively. Young children may have the correct pattern but a wrong choice of words. For example, saying "I wented" in place of the usual "I went" may actually be an indication of growth in language ability as it shows an awareness of the usual means of stating the past tense. Insisting that a child repeat the correct form does little good. If he should memorize the "I went" form, he will probably continue with "I goed, I camed, I seed, etc." until he has "internalized" the language understanding. After he has mastered this idea he will begin to note the exceptions. In learning to read by a language-experience approach, a similar awareness toward written language occurs. This is aided by appropriate remarks by the teacher as children watch her write. Exposure to this process enables them to learn phonic skills such as the various sounds associated with letters and how to use these sounds as an aid to recognizing words. They also learn the so-called comprehension skills such as understanding the main idea and noting a sequence of events. Ideally a "reading skill" is demonstrated when applied to reading. For example, when children can pronounce and state a meaning for words like *rake, hope, bike* and *fine* they have demonstrated an awareness of the "silent *e* rule" even though it was never directly taught and cannot be verbalized. Nothing would be gained by having them memorize this rule. Some children need many more experiences with words than others. They may be helped with planned exercises chosen from commercially prepared materials but preferably made by the teacher to meet a specific need. A child not quite sure of words like *hop* and *hope* might be given two lists of words that he knows as *hat, cap,* and *hate, cape* and helped to conclude that the *e* governs the vowel sound.

Ordinarily, learning phonic rules or verbalizing "reading skills" is not part of this program. The teacher needs to be aware of them in order to give intelligent help.

Evaluating progress and interpreting the program to parents

The problem of ranking pupils will probably never be settled satisfactorily. We vacillate between judging pupils in relation to group expectations and judging them according to their own expected progress. In an individualized program two questions always arise. The first is usually, how can you be sure there is progress when pupils read in so many different books especially when you're not even sure of the reading levels of the books? This is simple. Begin with a reading inventory and recheck periodically.[9]

The second question is more difficult to answer. How do you rank or assign a mark to a pupil who is reading above grade level but not putting forth too much effort? Conversely, how do you rank a pupil who is reading below grade level but working very hard? Ideally progress should be reported without letter or number grades. A realistic report either written or preferably oral in special conferences means more than some letter grade or check mark. All marks are subjective and often do more harm than good. In any event, the instructional level should not be the major criterion. A pupil whose instructional level is above his grade placement would not automatically receive an *A*. Since you will actually know more about each child in an individualized program it should not be any more difficult to assign a grade, where that is necessary, than under a traditional program.

Parents appreciate a frank discussion and direct answers to their questions. They often want to know how their child compares with others. In general, a pupil reads above, at or below a level usually expected for his age. Our school policy is to inform parents as specifically as we can when a child is below average. Obviously an explanation meaningful to one parent might not be uderstood by another. Some statements about ten-year-olds (5th grade) will serve as examples.

---

9 See Chapter 9.

David is still having a lot of trouble in reading. Although he is able to read some selections at fifth grade level it is a struggle because he has no method for figuring out new words. He is particularly confused about vowel sounds and is working on these in both reading and spelling. Most of the material he is using is on a third grade level . . . .

Nancy reads very well orally and can pronounce almost any word, but she has difficulty answering questions about her reading. This is true even with very easy books. She is being helped by reading to kindergarten children. She reads the story first, decides on a few important questions to ask them and then discusses the questions with me before meeting with the kindergarten children . . . .

You might compare a child's actual performance with his potential ability and emphasize his immediate instructional needs and tell what is being done to help him. Since every child is working at his instructional level, every child has some problem or immediate instructional need no matter how high his reading level is in relation to his grade placement. We usually speak of the needs in three areas—word recognition, comprehension and "other." The latter includes oral reading, work habits, creativity in project work, making personal use of reading, etc.[10] We sometimes predict rate of progress[11] and we always review the pupil's records[12] with the parents. They will see his list of books read, examples of work done and samples of follow-up activities.

Parents come to conferences with varying degrees of understanding. Two extreme views learned from their children are that no reading is taught at all or that reading is the only subject taught. To correct this, some teachers suggest that "English" or "Spelling" be written on creative writing papers and remind children when they are having reading. "You were learning to read to follow directions, but it was about social studies." Actually a visit to a classroom by a concerned parent is the best way to clear up this misunderstanding. Two other parental concerns should be mentioned.

---

[10] See Chapter 7.

---

[11] See Chapter 9.

---

[12] See Chapter 11.

Some parents are disturbed when a child concentrates his reading in one area such as horses, sports or mysteries. If you are unable to convince them that no lasting damage will result, suggest books related to other areas of the curriculum, world or local events, holidays or some other classroom project. Children will join interest groups to help others. Another approach is to suggest related nonfiction such as biography, history and travel. The child would gain considerable satisfaction by being considered the class expert in his chosen interest. Although this does not change his interest it might cause him to improve the quality of his reading. Using material like the Literature Sampler[13] helps everyone broaden his reading interests. Parents could spend more time with their children and help them by taking them on trips, discussing interesting topics at meal time, talking about their reading, going to book stores and libraries to browse through and buy or borrow books, joining a book club or subscribing to a children's magazine. Another more serious problem is the demand on a child to read more difficult books. In handling this situation, a teacher must explain the range of reading levels within a classroom and then review the child's level, his instructional needs and predicted rate of progress. The wide range of reading ability comes as a surprise to some parents and pupils. This is particularly true for those who have been in the top group of a basal program. The child and the parents must be reassured and must understand that progress is faster when the material read is easy and enjoyable. Let parents help when they ask, but be sure to suggest something they can do and something a child would enjoy. These activities might be related to reading such as visiting points of interest and then talking about the trip to enlarge a child's background of experience, reading and discussing stories to increase his listening vocabulary and help him with comprehension and word meaning, giving a child responsibilities around the house and offering praise for jobs well done, building a puppet theatre and showing an interest in puppet shows based on stories, encouraging children to dramatize stories, and including children in family discussions. Older children can read to help solve a family problem

---

[13] *The Literature Sampler.* Learning Materials, Inc. 600 Madison Ave., N.Y. 10022. See Chapter 1 for description.

Photo 4-11.

such as where to go on a vacation, how to make or do something, and choosing and taking care of a pet (Photo 4-11[14]). Parents can set an example by reading themselves to show that reading is a worthwhile activity and when children bring papers home from school offer praise and display them in a special place.

---

[14] Becky Hall, Kindergarten.

Continuing and extending an individualized reading program

The assumption so far has been that you will be the first to begin an individualized program with your class. When one of our pupils was asked how the individualized program compared with "the other way of teaching reading" she countered with, "Is there some other way?" What if your job is not to begin the program, but to continue it?

All read while the teacher determines instructional problems

An individualized program is no panacea (Photo 5-1). Each successive grade level brings a greater range of reading ability. The nature of the problems at eighth grade differ from those at third but there are just as many. Your first job is to discover these problems. The best way to do this is with an informal reading inventory.[1] Work with one child at a time while the rest of the class reads. You will note those who have difficulty settling down to read and take them first. Their problems must be attacked early if the class is not to be disrupted. It takes insight into the reading process and considerable practice to get reliable results from a reading inventory.

---

[1] See Chapter 9.

You might investigate the use of standardized *diagnostic* reading tests to supplement your findings.[2]

Some teachers new to the program like to combine the reading inventory with tests designed to check on basal readers.[3] Give a word recognition test[4] to locate the level where a pupil scores 100%. Then give the sub-tests at that level designed to assess word skills and phonics. If a pupil, regardless of grade placement, can pronounce correctly all the words at 3–2 level but makes errors at 4th level, he could be given certain parts of a 3–2 test. Those for the Betts Basic Readers[5] would be Part 2 (Consonant Sounds in Words), Part 3 (Vowel Sounds in Words), Part 4 (Number of Syllables), Part 5 (Accented Syllables) and Part 6 (Vowel Sounds in Accented Syllables). Then check comprehension with an informal inventory. Determine the highest level where the score is at least 75% and give the sub-tests that measure aspects of comprehension. The same pupil might score 100% at 2–1 level, 80% at 2–2 level and 50% at 3–1 level. He would be given the following parts of the 2–2 test: Part 9 (Finding the Right Part), Part 10 (Two Meanings), Part 11 (Finding What You Want to Know) and Part 12 (Thinking About a Story). The results of such tests may help you pinpoint a problem. Sometimes observations made during the testing are more helpful than the actual results and must not be overlooked. These include difficulty understanding directions, need for constant reassurance, apparent guessing, unsystematic work habits, working too slowly or too quickly, and emotional upsets.

---

[2] Alpert, Harvey et al. *Individual Reading Evaluation—Grades Primer Through Six.* Jericho, N.Y.: New Dimensions in Education, Inc.; Botel, Morton. *The Botel Reading Inventory.* Chicago: Follett Publishing Co.; Durrell, Donald D. *Durrell Analysis of Reading Difficulty.* N.Y.: Harcourt, Brace & World, Inc.; Gilmore, John V. *Gilmore Oral Reading Test.* N.Y.: Harcourt Brace & World, Inc.; Gray, William S. *The Gray Oral Reading Tests.* Indianapolis: Bobbs-Merrill Co., Inc.; Leavell, Ullin W. *Leavell Analytical Oral Reading Test.* Minneapolis: American Guidance Service, Inc.; Smith, Nila B. *Graded Selections for Informal Diagnosis.* N.Y.: New York Univ. Press, (Grades 4–6, 1963; Grades 1–3, 1966); Spache, George D. *The Diagnostic Reading Scales.* Monterey, California: California Test Bureau.

---

[3] Achievement tests accompany nearly every basal reader series. We use those published by American Book Company.

---

[4] See Chapter 9.

---

[5] Betts, Emmett A. and Welch, Carolyn M. *Betts Basic Readers.* N.Y.: American Book Co., 1963.

Photo 5-1.

Some teachers prefer to handle problems as they manifest themselves during pupil conferences. This is particularly workable for comprehension. You can be quite systematic. Make a list of comprehension "skills"[6] and check a few at a time during conferences. This may be done with one of the books the pupil has just read or you could give him a selection to read while you are checking over his records. Build up a file of short selections at various reading levels for this purpose. Remember that you are not after a mark or a grade. If an error is made, the important thing is to find out why.

Planning work on problems

Ideally each pupil should receive help at the time he realizes the need. If extended work is necessary, plan this with the pupil during a conference. Some pupils need more help and direction than others. They should receive it. An individualized program does not have to be non-directive. When a pupil appears to be avoiding reading by showing more interest in projects or having difficulty finding books you should try to determine the cause. It may be that you have a limited supply of books that would interest a very poor reader in an upper grade or the pupil may have personal problems that interfere with his learning.

We use a variety of techniques all aimed at making pupils

---

[6] See Chapter 10.

responsible for their own work. Worksheets, for example, are mounted on cardboard and the answers placed on the back. In second and sometimes third grade, the answer sheet is an exact replica of the worksheet. This is easier for young children to understand. In the upper grades the answers can be written or typed on the back. In both instances the answer sheets can be filed separately if you prefer. The pupils use these individually and in groups. They do an exercise and correct it immediately. They may work in pairs or groups of three. The procedure for groups is to have each pupil do the exercise alone, all check and agree on the answers and then check against the key. Another way is to have the group discuss the problems and decide on the answers and then check with the key. Pupils usually have a special form where they record their work. This may include date, sheet number, number of possible responses and number correct. It is reviewed by the teacher during the conference. Some teachers will question this procedure and suggest that some pupils will put forth little effort. This is particularly true when all written work is considered a test. Make a distinction between a learning activity and a test situation. Pupils help each other learn but each person is ultimately responsible for his own attainment (Photo 5-2).

Photo 5-2.

Keeping the approach interesting

Most pupils will continue to improve through reading and follow-up activities. Others may need specific work assigned for practice. Variety will add interest. Underlining words and filling in blanks are not the only ways of supplying answers. Set up folders for self-checking activities.[7] On the left, print the directions. Attach an envelope under the directions to hold pictures, letters, words or an answer sheet. Place the game or activity on the right inside of the folder. The folders can be filed in a cabinet or cardboard carton. Examples follow:

Long and Short *a* and *e*

Directions:
1. Take the words out of the envelope.
2. Say each word and decide which vowel sound you hear.
3. Place each word under the right column.
4. Check with a partner. The answers are on the backs of the words.

An envelope containing words like *bread, sea, gate* and *that* is glued under the directions. Each word has a colored dot on the back—purple if vowel is long *a,* red for short *a,* green for long *e* and blue for short *e.*

On the right are four columns. The headings are long *a,* short *a,* long *e,* short *e.* The appropriate dots are under each heading. After the pupil places the words, he turns the cards over. If all the dots are the right color, he has placed the words correctly.

Initial Consonant Sounds, Blends, or Consonant Clusters

Directions:
1. Remove the pictures from the envelope.
2. Place a picture below the letter that stands for its initial sound.
3. Check your work with a partner. When you turn the pictures over you will see the answer.

The envelope contains pictures cut from books and workbooks

---

[7] Some of these were suggested by Mary Brassard at an N. D. E. A. Institute held at Boston University.

and mounted on cardboard—cake, moon, fish, lion, violin, etc. On the back of each picture is the letter identifying the initial consonant sound. On the right of the folder are letters (t, h, f, d, g, l, etc.) with a space beneath each for a picture. When the pictures are turned over the letters should match (Photo 5-3).

Synonyms

Directions:
1. Read the column of words on the right.
2. Using the word cards in the envelope choose a synonym for each word.
3. Check with a partner. The answers are on the backs of the cards.

To distinguish the answers from the task, print the column of words and the words to be matched in some color (green) and the answers in another color (blue). A word in the column might be *pretty* printed in green. The matching card will be *beautiful* also in green. *Pretty* will appear on the back of *beautiful* in small blue print (Photo 5-4).

Photo 5-3.   **Photo 5-4.**

To Check on Various Rules (If that is important)

Directions:
1. Read the rule. (Example—when a one syllable word ends in one vowel followed by a consonant, double the consonant before adding endings such as *ed, er, est,* or *ing.*
2. Read the words. Decide whether or not the words follow the rule. If you have a partner, talk about each word before you decide.
3. Take the *Yes* and *No* cards from the envelope. If the word follows the rule put a *Yes* card by it. If it doesn't, put a *No* card by it.
4. The answers are in the envelope.

Another way of doing this would be to have words in the envelope and two columns headed *Yes* and *No*.

Root Words

Directions:
1. Work with a partner.
2. Decide what the root word is.
3. Write it on a paper.
4. Check your answers. See the envelope.

Fifty words like *shady, famous, astonishment,* etc. are typed on colored paper and pasted in the folder. The answers are in an envelope also pasted to the folder.

Locating Information

Directions: (Do with a friend.)
1. Take the pictures of the books out of the envelope.
2. Read the questions on the folder. Talk about the answers with your friend and decide which "book" would be best.
3. Put the right "book" next to the question.
4. Check your work. The answers are on the backs of the "books."

5. Which book would you like to read?
6. Find one like it in your room. After you read it, tell your friend about it.

Cut out pictures of books from catalogues. Make up questions to go with the books like, "In which book would you look to find out what fish eat?"

If you think a pupil would benefit from writing words, use exercises like the following:

Classification—On colored cardboard print two words under which others could be classified such as (1) country (2) city. You might want a third classification (3) both. List words like *pigs, skyscraper, barn, sidewalk, hay* and have the child write the words in the correct column. Place the correct responses on the back of the card. Older pupils can be given words or statements and asked to group them and provide a heading.

Comprehension and Vocabulary—Cut colorful pictures from mail order catalogues. Mount them on cardboard and provide questions on vocabulary and comprehension. End with a personal question such as, why would boys want to buy this toy?

The best type of exercise is related to the child's reading. Develop and keep on hand ideas that can be adapted to any book.[8]

1. Some words indicate who, when or where. Find five words in the story you just read to illustrate each of these words.
2. Every time you come upon a word that tells how someone moved write it on the board or put it on a card and pin it in a special place on the bulletin board.

---

[8] Sources of ideas for learning activities are: *Educational Games and Activities* and *Games and Activities for Early Childhood Education*. (Teachers Publishing Corp., Darien, Connecticut); *Spice, Probe, Plus, Spark, Create, Action, Stage, Rescue*. (Educational Service, Inc., P.O. Box 219, Stevensville, Michigan, 49127); *The Other Children* (Harper & Row); *Helping Young Children Develop Language Skills: A Book of Activities*. (Council for Exceptional Children, 1411 South Jefferson Davis Highway, Suite 900, Arlington, Virginia, 22202); *The Headstart Books: Looking and Listening, Knowing and Naming, Thinking and Imagining*. (McGraw-Hill Book Co.); *Reading Games That Teach*. (Creative Teaching Press, 514 Hermosa Vista Ave., Monterey Park, California, 91754) and *Learning Activities from British Schools*. (Leicestershire Learning Systems, Box 335, New Gloucester, Maine 04260).

3. Write down all the words on these two pages that begin with blends. Check them with a friend and then say them to me.
4. In the book you just read, find five factual statements and five opinions.
5. For a biography, travel or other factual type book, what are some questions that were not answered? Where might you get information to answer these questions?
6. Write one question about your book beginning with each of the following words—who, what, when, where, why.

Poor readers in the upper grades will enjoy making exercises and games to be used with younger children. This is not only a more respectable activity but actively involves them in the learning process. If they use their materials with younger children and with each other, they will get more meaningful practice.

These ideas have been offered to suggest a variety of ways to help pupils who are experiencing difficulty in reading. They should be used with discretion. Providing ample time for reading and creative writing will enable most pupils to learn without artificially imposed exercises and worksheets.

Reading to make things

All reading does not have to consist of stories and articles. Provide directions and materials for a variety of projects. Since the objectives are to teach reading, to follow written directions and to understand diagrams, the project itself should take relatively little time. Different activities can be added throughout the year. You might have the directions filed on cards and the materials in a scrap box. A more orderly arrangement is to place the materials necessary for a specific project in a shoe box and put the directions on the inside of the lid. See Photo 5-5.

Children will adapt ideas from projects to illustrate books. The possibilities are endless: different types of puppets, pin cushions, decorated boxes, book covers, mobiles, paperweights, greeting cards, book covers, ornaments, games, doll furniture, paper flowers and jewelry. Pupils enjoy paper folding and can use the items they make to illustrate stories (Photo 5-6).

Photo 5-5.   **Photo 5-6.**

MAKE AN ALBUM OF LEAVES

You will need:

white paper
crayons (mostly green)
scissors
paste
leaves

1. Put a leaf on your desk or table. If you have the underside up, it will print better.

2. Put a piece of white paper over the leaf.

3. Choose a crayon as nearly the same color as the leaf as you can find and while holding the paper firmly color over the leaf.

4. Cut out the leaf.

5. Paste the leaf on a piece of paper.

6. Do the same thing with other leaves.

7. Place your pages between two blank sheets of paper and staple them all together to make a book.

8. Decorate the cover of your book.

Your album will be more interesting if you label the leaves and write something about them.

Now that you know how to make a leaf album you could write directions for making a "coin collection" or some other album.

MAKE A "FOSSIL"

You will need:

modeling clay (or wet sand)
cardboard
paper plate and wax paper (or just a hole in the ground)
shell (or twig, leaf, insect)
water
plaster of Paris
water color paint if you wish

1. Roll out some modeling clay until it is smooth.
2. Place a shell on the clay and put a piece of cardboard over it and press down firmly.
3. Carefully remove the shell.
4. Place the clay model in a paper plate that is covered with wax paper.
5. Take about a quarter of a cup of water and slowly add plaster of Paris to it until it is quite thick. You will have to keep stirring it.
6. Pour this mixture over the clay model.
7. When the plaster of Paris dries remove the clay and you will have a model of a shell fossil.

Your "fossil" will look like a real one if you paint it the color of earth. Use water colors.

You could make a collection of "fossils" and put them on display. If you do, label them so that people will understand your display.

Different organizational plans for junior high

As individualized reading is extended into junior high school problems may arise if reading and English are taught by different teachers. A better way is to have one teacher assigned to the same group of pupils for two consecutive periods—a class in Language Arts. This makes it possible to continue a somewhat flexible program. During this language arts block pupils read, discuss and write. Sometimes the whole period will be used for reading or for writing or practicing a play. At other times a class might work for weeks in one area like biography, dramatics or creative writing.

English and reading teachers at the junior high level tend to be more specialized in their interests and abilities than elementary teachers. To take advantage of this, a non-graded approach is being used in our junior high. All pupils (about 700 in grades seven and eight) regardles of grade or ability sign up for language arts classes by interest. They change four times a year and could have four different teachers a year. They meet in double periods five days a week. The only real difference is that each class has a definite center of interest. The approach is still individualized. This, in fact, becomes more necessary than ever because of the wider range of ability. Pupils choose their own reading matter related in some way to the general topic. If the topic is mystery stories, most will read mysteries but some will take an interest in background material or biographies of authors. Teachers are enthusiastic as classes are never taught the same way twice.

Pupils choose from the following:[9]

1. *Journalism*—Make your own class newspaper, visit a newspaper plant, watch movies! Using your own daily paper, you will take a close look at journalism as part of a democracy. You will study parts of newspapers and magazines, make comic strips and cartoons, write news articles, interview interesting people, take field trips and discuss current events. All cub reporters report to journalism class!

2. *Sports*—The thump of a football, the flashing of ice skates, the crack of a baseball on a bat; all this and more. Explore the world of sports through books, magazines, newspapers, television and film. Your activities will include sportscasting, writing reviews and editorials, interviewing sports figures, sports cartooning and discussions of current sports events. Put yourself in the place of Bobby Orr or Joe Namath in this exciting fast-paced course!

3. *Creative Writing*—To be or not to be an author! This could be your aim. Writing can be fun and can open your door to fame. Would you like to write a novel, a play, or a short story someday? From one sentence beginnings through narratives,

---

[9] Course descriptions were written under the direction of Gary Lisherness, Chairman of the Language Arts Department, by the following: Robert Brown, Helen Gallagher, Ernest Gauer, Anna Hutchinson, Ruth Niles, James O'Callaghan and Elizabeth Poulin.

expositions, and short stories, this course will progress as you learn to express your ideas on paper. Get an early start . . . sign up now for creative writing.

4. *Myths and Folklore*—Ride on the winged chariot of the sun, kill the monster Grendel with your bare hands, fight alongside of Lancelot in the forefront of King Arthur's army! Explore legends that have stirred the imagination and blood of young people for the last two thousand years. Live again in your mind the deeds of might and valor, of heroines and heros, of traitors and patriots. This course offers a stimulating and entertaining adventure in reading and writing.

5. *Mysteries*—Who dunnit? Was it the butler, the maid or the redhead? In this action-packed course, you will meet the greatest spellbinders of the mystery world. Chills will run up and down your spine as you try to solve these mysteries. The flash of a knife in an alley, a scream in the dark, the ghostly shadow flitting away! Plan your own mystery story! Pick this course if you think you can stand the excitement.

6. *Short Stories*—How short is a short story? How long is a short story? The world of mystery, adventure, and humor is waiting for you! You will read short stories of your choice, discuss and dramatize them. You will study the authors to learn how their lives influenced their writings and then . . . you will be ready to write your own short story. Some of them may sell—who knows?

7. *Man and His World*—WHY? HOW? WHEN? Are you interested in what's going on in the world? Do you know what the problems are and what's being done about them? This course will offer activities using newspapers, magazines, books, films, field trips and local speakers to show you how everyday events influence our lives. Find out what you can do to make our world a better place!

8. *Biography*—Will your biography be written and published someday? In this course, you will read about the people who interest you. They may be inventors, explorers, athletes, political leaders, anyone you would like to know. We may read from different authors to develop critical viewpoints. You will have a chance to write a biography or even your auto-biography.

9. *Development of Our Language*—What is the origin of your

name? Do you know the history of our language? In this course, you will learn how the English language has developed and how it changes. You will see how the linguist views grammar and will have an opportunity to improve your own writing and speaking.

10. *How to Study and Do Research*—Become adept at using the works of others to your advantage! Make the library work for you. Interview well-known public figures and unknown little characters who escape public attention. Become skilled at researcher's "shorthand" and delight in your ability to expand that "shorthand" into something exciting and readable.

11. *Oral Communication*—A good story teller is a popular person. He knows just what to say and when to say it. You can be a good storyteller! You can learn to emcee a banquet, win a debate or preside over a club meeting; discuss a television program, movie, trip or game; paint a vivid picture with words; or make puppets and put on a show.

12. *Poetry*—Writing lyrics for your guitar, limericks, jingles, serious verse, blank verse are just some of the activities planned for this class. Your skill will develop as you hear, read and speak works of master poets and become acquainted with rhyme, rhythm, figures of speech and types of poetry. There will be recordings, tapes, filmstrips and movies to see, hear and make!

13. *Famous Authors*—Famous authors of the present and past will come alive to you as you become acquainted with their works and the details of their lives. Take an imaginary trip with Mark Twain down the Mississippi River and a real trip with your teachers and classmates through the woods surrounding Thoreau's Walden Pond. A good imagination and a little "Yankee Ingenuity" will help you to make your book friends live again for your classmates, and they, in turn, will help you meet famous people.

14. *Play Production*—If you would like to become an actor, or director, or work back stage painting scenery, doing make-up, or handling lighting of the theatre, then you *must* join us on stage. We will read about the history of the theatre, learn about the parts of the stage, do charades and pantomimes, perform radio plays, and put on a one-act play. Join up! This course offers fun for everyone.

15. *Radio and Television*—Learn the skills of audio-visual edu-

cation. Visit a radio station, tape your own program, make film strips and produce a TV show. Script writing and speaking and acting skills will be developed.

16. *American Folklore and Legends*—Who is your favorite American hero? If you like the imaginary heroes like Paul Bunyan and John Henry, you will enjoy this course. You will read, listen to recordings, and view filmstrips. We'll even use music and movies to tune you in on characters like Casey Jones, Johnny Appleseed and Sweet Betsy from Pike! You will write your own folk tale or legend.

17. *The Novel*—This course offers an inexpensive ticket to other times and other places. You will choose your own novels to read. After learning about plot, character and setting we will discover together how they relate to each individual story. We will study the structure and skills necessary for relating oneself to experiences of others.

What happens in these courses? The basic pattern is reading with individual conferences, group sharing of books, focus on a specific topic followed by creative writing. The specific topics are sequenced or related in a way that leads to a satisfying culminating activity. Those who choose mysteries might work on opening sentences, mysterious words, building suspense, and character descriptions. Those interested in poetry might focus on rhyme, rhythm, similes, alliteration, or types of poems as haiku or such topics as biographies of a favorite poet, geographical locations described in poems or the historical background. All of the topics can be made interesting and rewarding to all pupils regardless of ability or reading, writing or spelling levels. A brief description of a journalism course[10] will serve as an example.

Books are collected in the classroom on all aspects of publishing —famous journalists, publishing businesses, printing, reporting, writing news items, making cartoons, advertising, photography, vocations, and anything else related to journalism. Numerous newspapers and magazines are added. These, along with library resources and a copy of the local daily newspaper for each pupil, form the core of the individualized reading program. As in all our classes the pupil-teacher conference is important in this phase. Newspapers

---

[10] Developed and taught by Gary Lisherness.

lend themselves to the development of many comprehension skills —skimming, scanning, identifying the main idea, noting details and sequences of events, recognizing propaganda and inferences, comparing headlines, articles and editorials for local or political slanting, appreciating humor and satire in cartoons and locating specific information. The local newspaper is literally cut up as pupils examine its different parts. Both general and technical vocabulary grow through wide reading, films and discussion. Crossword puzzles are a natural for word meanings and spelling.

Articles from the daily paper may be rewritten to reflect new information or summarized for a weekly. Pupils interview people outside of school and write feature articles. About half way through the course a trip is taken to a newspaper plant and the most stimulating activity begins—making a newspaper. The class divides itself into many groups each responsible for a section of the paper. There is a place for everyone—editorials, world and local news, sports, music, poetry, women's fashions, cartoons, comics, advertisements, etc. In addition, everyone has a job—editor, feature writers, reporters, proofreaders, typists, printers, distributors. The result is a very sophisticated newspaper for a junior high. One boy[11] was inspired to write a poem.

HEADLINES

Mets take flag;
Dodgers sag.
Plane crash kills ten,
Burglar strikes again.
Man steps on moon;
Japan hit by typhoon.
Final clearance sale;
Harbor Patrol sights whale.
"Giant" guard breaks thumb;
Police arrest drunken bum.
Congress passes new law;
Plan goes without a flaw.
Rams win on ninety yard pass;
Miners discover natural gas.
Judge takes side of the *New York Times*
Everything in this poem rhymes.

---

[11] Peter Westervelt.

6

Using an individualized approach with children who have learning problems

Learning to read for some pupils is very difficult. Studies of failures have revealed many causes, but for every child who fails there are others in similar circumstances who succeed. Numerous tests are used to try to predict readiness for reading and considerable time is spent preparing children to do well on these tests. Unfortunately none are completely reliable. Some children who do well on the tests fail to learn to read while others who do poorly are labeled failures before they have an opportunity to try. You will not have these problems when you use a language-experience approach. All will not learn with equal rapidity but everyone will experience success. Language understandings can be presented and assessed in a natural way and more experiences can be provided as needed. Because the program is flexible individual learning problems can be handled in the classroom without making the child feel different. This should be obvious in the programs that have been described so far, but because questions are often raised about atypical pupils a brief description of their activities will be offered.

Ideally every child in trouble should have a complete diagnostic evaluation. Since most school systems do not provide this service these pupils will get no help unless some classroom teacher is able to give it. The following are suggestions for recognizing problems in classrooms. Some problems are actually a result of the school situation and are, therefore, the most amenable to change. The purpose of this chapter is to show how children with problems are being taught to read. The fast learner will be included not because he had difficulty learning but because he often becomes a problem in a traditional classroom setting. Thus, five broad categories will be considered: the slow learner, the fast learner, the emotionally disturbed, those with learning disabilities and those who for various reasons arrive in the upper grades unable to read.

The slow learner

Who is the slow learner in your classroom? If you think of the whole class, it's a matter of relative standing but there are some children who would be considered slow in any class. (Always remember, though, that even your slowest pupil learned to talk without any formal schooling.) In an individualized program everyone can contribute to all activities. Regardless of the subject, there are reading materials at all grade levels. This is particularly helpful in building self confidence and maintaining a good self concept. The main difficulty from an instructional point of view is being able to accept slow progress. A good assessment with a reading inventory will tell you what to expect.[1] A slow learner will not gain a year's work in a year so this is an unrealistic goal. You may have to give more help in selecting books and suggesting follow-up activities. Since a slow learner's attention span is often limited you may have to offer many short selections and suggest projects that can be completed in a relatively brief period of time. You may be amazed at how long your pupils will stick to an activity that has real interest for them.

All teachers know how difficult it is to find a basal reader for the lowest reading group in the room. The choice is usually a compromise that interests no one and does not solve anyone's reading problem. Even more of a disadvantage, though, is the stigma felt by the pupils. In an individualized approach, each pupil gets the specific help he needs. This coupled with a remarkable change in attitude brings success. A girl[2] entered second grade in November unable to

---

[1] See Chapter 9.

---

[2] To avoid possible future embarrassment, no pupils will be identified by name in this chapter.

read. She recognized only four words from pre-primers of the three different basal programs she had according to her records. Six months later her word recognition scores on a reading inventory were 100% at pre-primer level, 100% at primer (1–1) and 72% at first (1–2). During this time she was in a language experience approach and had not seen any basal readers. Three boys who had spent two years in a first grade program began second grade in pre-primers. They were tested in November previous to changing to a language experience approach and tested again six months later in May. Their word recognition scores were:

	NOVEMBER	MAY
First boy	PP—40%	PP—100%
		1–1—100%
		1–2— 58%
Second boy	PP—50%	PP—100%
		1–1—100%
		1–2— 64%
Third boy	PP—80%	PP—100%
	1–1—35%	1–1—100%
		1–2— 96%
		2–1— 56%

Comprehension could not be checked in November but in May there were no comprehension problems in the comparable selections. The change in the attitude of these children toward learning was remarkable and was far more important than the gains in test scores.

When a sixth grade teacher[3] had a class write about favorite books a slow learner wrote:

> One day I was walking to the school library. I got a book—THE PINCH HITTER.[4] The story started and all of a sudden I was in the second chapter and soon I was threw the whole book and then I made a picture.

When classes are assembled heterogeneously, as ours are, there will be slow learners in every room. At first grade level, they will

[3] Linda MacLeod.

[4] Friendlich, D. *Pinch Hitter*. N.Y.: Westminster Press, 1965.

be dictating experience stories long after their classmates are writing their own stories and reading books (Photo 6-1).

At second grade level, they will still be dictating group charts and writing stories which they will read (Photos 6-2, 6-3, 6-4, and 6-5). As the year progresses they will gradually begin reading books.

By third grade, most slow learners will be reading. An observer would not be able to pick them out unless he was familiar with the reading levels of the books.

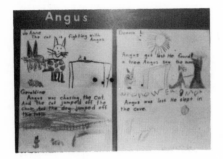

Photo 6-1.

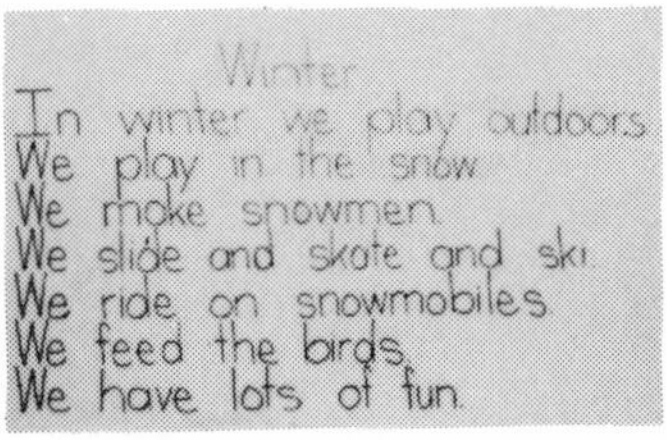

Photo 6-2.

Photo 6-3.

Photo 6-4.

Because a second grade class was upset when a boy lost his apple the teacher[5] suggested that they write a story called "The Missing

---

[5] Lucille Winters.

Apple." The following was written by one of the slow learners.[6]

> One day a little boy and girl was picking some Apple they pick all the Apple but two. They said lets save the two Apple. One nith they was two girls. They said I am hungry. The girls was running down the Road. They was tired. Then they saw the Apple trees. They said that was a nuff for us. They made a noies. A big noies. They woken the boy and girl. They told the father. He came out the Door with a gun. He shoot up in the air. He called the polices. The polices came to the house. The polices scare the two girls away. The End of the story. Good nith.

When we speak of reading levels we imply that a person reads everything at some one level. For most people, this not true. The more information you acquire about a topic the easier it is to read about it. Thus, you might be able to read very difficult material about antiques but have to read science on an elementary level. Encourage pupils, particularly slow learners, to become specialists in something. This doesn't mean that they spend all their time with one topic, but that they read on it from time to time and are recognized as experts by other members of the class. Sometimes an interest develops in relation to mathematics, science or social studies. One boy in sixth grade, for example, became an expert on Buddha and his teachings. Many pupils pursue their hobbies—sports, stamps, racing cars, fishing, model rockets, music, etc. Even a slow learner can know more about his specialty than the teacher. Slow learners should be included in all class activities. If given a chance they may prove to be your most creative pupils.

The fast learner

Fast learners in traditional classrooms have one common problem—boredom. While a slow learner is bored because of frustration a fast learner is bored with repetition and drill. One boy read his basal reader the first day while the teacher was getting organized. These are the pupils who get extra work, usually equally boring,

---

[6] Compare this version with those of the fast learner, the disturbed child and the pupil with a learning disability.

and sometimes develop poor study habits and become discipline problems because nothing is a challenge. In an individualized program, they are challenged because they read at their instructional levels just as everyone else does (Photo 6-6).

The teacher needs to show an interest in their work, encourage them in their varied pursuits and be willing to let them go on their own. It is not unusual to find fast and slow learners working cooperatively on reading projects by choice. Pupils respect each other for their contributions; artificial distinctions are eliminated. As a second grade girl reported;

> "Nathan and I done a TV show. It was about the new boy.[7] When we were done drawing we asted S to help us. I read about it and Nathan and S held up the pichers. S made a cupl of mustaks but that was OK."

One of the fast learners in the second grade mentioned above had this story about the missing apple:

> In the middle of the night a bank robber named Carl broke into the museum. There were two guards. Carl had his sleeping dart gun and he shot one of the darts at each of them. POW! POW! went the darts. After he had made sure they were sound asleep, he went into the room in which he saw what he was looking for. It was the silver apple. Being careful not to drop the silver apple, he ran to the window where he had come from and climbed out. He took the apple to his hideout in the woods. He called the city hall and the mayor answered.
>
> "I have the silver apple," said Carl. "You bring $1000 in cash and I will give it to you."
>
> "Where shall I meet you?" asked the mayor.
>
> "At the first maple tree you come to."
>
> "And what time shall I meet you there?"
> "At 5:30 p.m.," said Carl.
>
> And he did.
> Carl was a good man after that.
> Can you belive it?

---

[7] Justus, M. *The New Boy in School*. N.Y.: Hastings, 1963.

Fast learners can also pursue topics in depth. They need encouragement and recognition. They may be natural leaders for dramatics and other unsupervised group activities.

Photo 6-5.

Photo 6-6.

The child with emotional problems

Surveys indicate that at least ten per cent of the school population is emotionally disturbed. Such pupils have personal problems so severe that learning is inhibited. They may sit quietly and daydream or they may find it impossible to sit still. They may avoid doing their work or they may do it all but fail to remember it. They are usually in trouble with their classmates, either aggressively so or by being rejected. They generally are pupils of normal intelligence otherwise capable of learning.

No reading program will solve emotional problems. If no help is available in school, parents could be encouraged to seek help from other agencies. In the meanwhile, the child needs help in your classroom. It is a difficult situation because it often requires different standards for behavior and a degree of permissiveness unacceptable to some teachers. An individualized program allows this permissiveness and extra personal attention. Since children are free to move about they don't find themselves gaining attention by upsetting the class. Because the schedule is flexible they can continue with an interesting activity. If more direction and "structure" is needed, it

can be given to them just as other pupils are given different assignments. The individual conference makes it possible for these pupils
to have the complete attention of the teacher which some of them
crave. Some children are helped by reading about people with
similar problems and such stories can be made available.[8] Occasionally emotionally disturbed children are exceptionally good
readers as reading becomes a means of escape. One of the boys in
our special room for the emotionally disturbed is nine years old
and reads at seventh grade level. Some children manage to learn as
evidenced by test results, but are unable to do the class assignment.
They may score high on achievement tests but be given failing
marks by the classroom teacher.

A disturbed child who reads above grade level wrote this way
about the missing apple, rambling on for nine pages:

> Once apoun of time ling a ling a go there was a apple. But it
> was a yellowll apple. But it was so yellowll it looked like a little
> gold apple. But there was people buck then. But they did not now
> aboat the little yellowll apple thet was so yellowll it looked lake a
> lattle gald apple. But how showed the pleople now aboat the lettle
> epple that was a yellowll apple that look like a gold apple an last
> they go for a wake becaus the apple was in the woulds but one day
> a little girl and a little boy want for a walk in the woods and then
> at the same time the girl and the boy haller out look a lottle gold
> opple. And so the boy and the girl run for it. But the girl got the
> apple then she and the boy run home and told there mother and
> father I gold apple yellowll we must find out were the apple came
> (Photo 6-7) so way dont we have a little pisse of the apple then
> John and Lore find and they that thet the apple was a gold Apple
> But it is not your flgt you and Lore that thet it was a gold because
> the little apple was a panted little apple.

Teachers in an individualized program for the first time are sometimes overwhelmed with all the problems they discover. Instead of
being dismayed, be glad that you have identified a problem so that
help can be obtained. Unfortunately children just do not outgrow
their problems (Photo 6-8).

---

[8] Any children's librarian can supply you with titles of books about most situations. In addition to these, books on "Psychologically Relevant Themes" (*The
Man of The House, All Alone with Daddy, One Little Girl, My Grandpa Died
Today, The Boy with a Problem, Don't Worry, Dear*) are available from Be-
havioral Publications, Inc. 2852 Broadway, New York, N.Y.   10025.

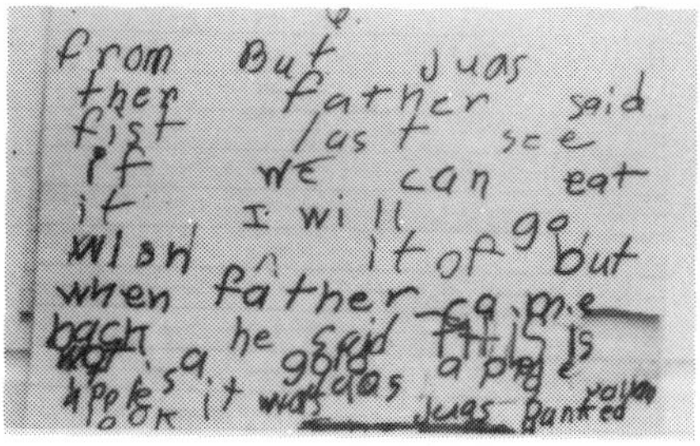

Photo 6-7.

Photo 6-8.

The learning disabled

These are the most misunderstood and neglected children in the classroom (Photo 6-9). Teachers accuse them of being stubborn, egocentric, uncooperative, lazy and messy. They are particularly annoying because they fail to learn yet have normal or even superior intelligence. It will be suggested that they have some peculiar hearing or vision problem but this will not be substantiated. Because there is no overt reason for their failure they make teachers feel inadequate and they upset their parents. These anxieties are somehow conveyed to the pupil who reacts by avoiding the learning situation as much as possible. You will note that these pupils are generally poorly organized. They are slow to begin their work and slow to finish. In fact, they may be so easily distracted that they have difficulty completing any assignment. They may be temperamentally unstable, have poor motor control and be confused with left and right directions. Often they show no time sense. They sometimes have a general language disorder—delayed speech, confusion in pronouncing words or an articulation defect. They may be slow in responding but they will give you the feeling that they are thinking of what they are going to say. When they begin reading you will discover that they cannot generalize from one situation to another, do not profit from context clues, have difficulty remembering words even from one line to the next and will read word by word. They may even forget everything during a vacation period. They will have great difficulty with spelling and may always have a spelling problem.

It takes an enormous amount of patience to work with these pupils. If you find that you are not able to teach them, seek help because the longer the child is allowed to fail, the more difficult it will be to help.

One such boy in the second grade mentioned above gave this version of "The Missing Apple." Notice that the ideas and the sentence structure are far superior to that of the slow learner.

> a lettle doy pit a
> Apple and he lan it at day
> He look and look and look
> But He bid not fiad it
> and He look in his toy box
> But He bidit fiad it sow He
> look in his bros but He
> didt not fiad it and sen He
> got a god Ider He wet out
> in to the Apple felb
> He pit a Apple sen He
> wet in to the homs and
> pot it dome and sen
> He wet to bed but He
> did not rele go to slep
> He woh the apple and
> all a sun He saw it
> go dome and dome He saw it go
> in a sekrit hab

Translation:

A little boy picked an apple and he lost it that day. He looked and looked and looked but he did not find it and he looked in his toy box but he didn't find it so he looked in his desk but he didn't find it and then he got a good idea he went out into the apple field he picked an apple then he went into the house and put it down and then he went to bed but he did not really go to sleep. He watched the apple and all of a sudden he saw it go down and down, he saw it go in a secret hole.

These pupils have been rediscovered in the last few years and all kinds of teaching suggestions and materials, old and new, are available to teach them. We have not used any unusual techniques with our pupils. We intervene as early as possible and use a language experience approach but we do have one variation. We consider their words[9] reading words rather than just writing words so we reteach the words they forget rather than throw them away. We use the Fernald[10] approach to teach words not so much because we think words are learned by tracing and feeling them but because tracing forces the pupil to look at the word. Whatever the reason though this approach does work (Photo 6-10[11]).

The boy whose story appeared is included in all class activities. He has progressed from being a complete non-reader when he came in the beginning of the year to acquiring a reading vocabulary of about two hundred words. After three years of failure, this is quite an accomplishment. The words from his file box, kept in alphabetical order, begin with: air, astronaut, around, author, ball, bass, beat, been, boat, Bobby, bottle, box. One teacher[12] found that posters and charts were popular because they could be made in a relatively short period of time and then displayed—instant success. (See Photos 6-11, 6-12, 6-13, 6-14 and 6-15).[13]

The non-reader in upper grades

Why do we have non-readers in the upper grades? The pupils themselves have various explanations—"I was sick a lot in kindergarten. Even though I don't wear glasses I have something wrong with my eyes, How can I read a fifth grade book when every year they give me baby books to read, My mother can't read and I take after her, I read all right in the other school, I fooled around a lot in first grade." Many times the cause is difficult to trace. It may be that a real learning disability exists. Too often it appears to be related to grouping practices that keep some pupils at frustration levels. Each year the poor reader gets further and further behind. These pupils need more instructional time than most upper grade

---

[9] See Chapter 4.

[10] Fernald, Grace. *Remedial Techniques in Basic School Subjects.* N.Y.: McGraw-Hill, 1943.

[11] First grade girl learns the word, *what.*

[12] Julia W. Kierstead.

[13] Pictures illustrate the work of two boys and a girl.

teachers can give them. Furthermore, few upper grade teachers know how to teach beginning reading. Thus, the non-readers are grouped with the poor readers and all are either placed in a book designed for much younger children or are subjected to the same type of word and phonic drills that have consistently failed them. One such pupil arrived in a fifth grade unable to recognize much more than his own name. He couldn't name all letters of the alphabet. Considerable testing revealed normal intelligence and no specific learning disability. He was completely defeated though and at first just sat and cried. A year and a half later, with the help of two inexperienced but understanding teachers,[14] he was reading at second grade level and writing stories.

> I have a 1970 Chevy. It is white with Dual tires, and a cylone on the hod. I enter in the Indianopolis 500. I had to think of a number for my car and I put down number two. Then I try it out on the race car track. It is 2 miles long. I made it in a minute. I was readey for the race and I could not wait. And the day was here and I saw all kinds of car ther. We have to start at the startoring line. He said go. Number 44 was ahead. I was going fast. I was ahead. Then I had a flat tire. I had to go to the pit. Then I got my tire fix and I was in the race again. Number 44 was in trub. His car was out of control. He was hading for the fence and it happened his car was a total reck. He was rush to the hospital. And there was only 4 car left. I was in second place but I was move up so I was ahead. . . . I won $5000. I was so pro that I kiss my car.

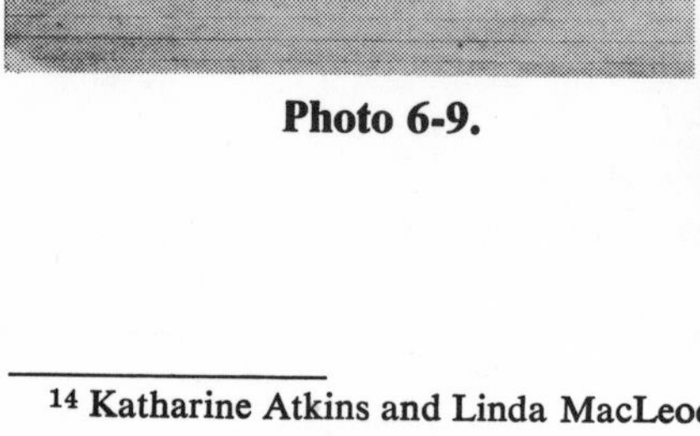

Photo 6-9.

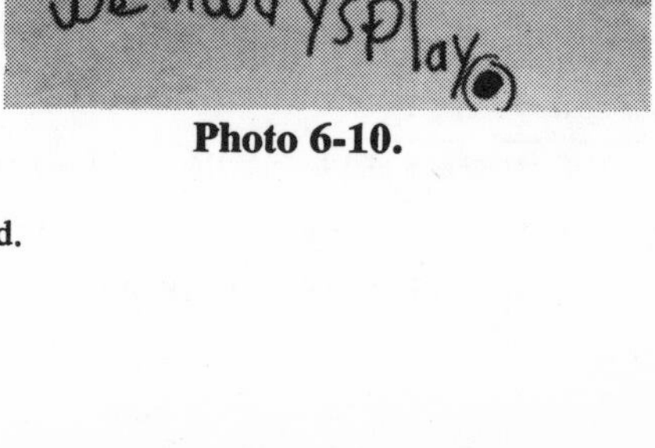

Photo 6-10.

---

[14] Katharine Atkins and Linda MacLeod.

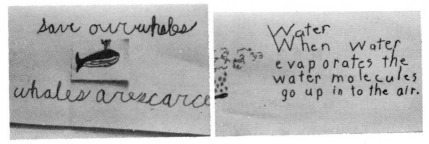

Photo 6-11.   Photo 6-12.

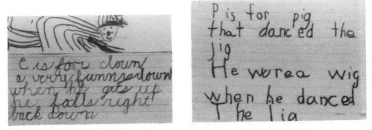

Photo 6-13.   Photo 6-14.

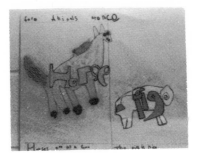

Photo 6-15.

Photo 6-16.

Their own dictated and written stories are the best source of reading. When they are ready introduce the Reader's Digest material and books published by Benefic Press and Field Educational Publications[15] as well as selected short items from newspapers and magazines. Children's stories can be read for the purpose of reading them to non-readers in kindergarten.

Teachers in individualized programs are acutely aware of pupils' problems. Because a thorough analysis is made, causes can be determined and realistic goals set. Pupils can be given varying degrees of help in a manner most appropriate to their learning style (Photo 6-16). The flexible schedule and accepting atmosphere of the classroom promote feelings of worth and well-being—two prerequisites for learning. Above all, such pupils are able to contribute to all activities without fear of censor by their classmates.

---

[15] See Chapter 1.

Stimulating pupils to be creative in follow-up activities

When you read a book you usually want to tell someone about it. Children do too. It is unnecessary and would be impossible to have every child report on every book read. Teachers used to the basal reader approach are sometimes disturbed because they cannot ask detailed questions over every page. This may be necessary in a basal program because it takes three days for one story. It is a group lesson though, so one child is not expected to answer all the questions. In an individualized approach, teachers are aware of each pupil's needs and can give consistent help, even from day to day if desired, from whatever the child chooses to read. Teachers who do not realize this difference in the two programs often begin by requiring elaborate written reports over every book read. Nothing could be more detrimental than this. Book reports, written or oral, are generally boring and uncreative. Our approach is to encourage pupils to react personally to their reading. They share with their friends, the teacher, the whole class or with other classes. They stimulate each other and when left alone, children are extremely creative. Even first and second graders write and produce their own plays and puppet shows (Photo 7-1).

Danette, Kathy and I made a puppet show. It is The Big Whistle.[1] Kathy made the little boy Mitch and the mother holding

---

[1] Beim, Jerrold. *The Big Whistle*. N.Y.: The Macmillan Co., 1968.

a lunch box. Danette made the bus. I made the father. Kathy thought of it and asked me to read the book so I could be in it. I sayed yes. We had a good puppet show.[2]

The follow-up activities suggested might be found in any type of classroom. The difference is that in a classroom designed for individualized reading they are a recognized part of the instructional program and may be the basis for spelling, creative writing, English, dramatics and the various reading skills. They bring children together and are entirely pupil directed. The teacher makes suggestions, sees that materials are available and schedules time for sharing, but generally is free to teach individuals and small groups.

Photo 7-1.

Activities pupils find rewarding

What do you do after you read a book? Sometimes you just read another one but other times you paint a picture, write about your favorite character, make something or do something else. What did you enjoy doing most this year?

---

[2] Bridget Hallee.

Second Graders

We made a play. It was about a house that was very busy.[3] I had never been in such a nice play. It took a long time but we did it.[4]

Kim and Danette and me made this puppet show. It was so good the teacher let us put it on for almost our hole wing. We made our puppets of otag paper. First in the story[5] Patty wanted a pet so her daddy took her to the pet shop.[6]

Me and Roxanne made a prodget out of clay.[7] Roxanne made the boys and I made the girls. It was fun.[8]

We made illustrations then the teacher wrote the name of the book on a piece of paper. And we had to guess. I made a illustration about Olly's Alligator[9] (Photo 7-2). We had fun doing it. But we had a hard time doing it.[10]

We made a TV show.[11] "I told about it and Katy and Paula rolled it up.[12]

---

[3] Weigle, Oscar A. *The Busy House.* N.Y.: Wonder Books, 1959.

---

[4] David O'Connor.

---

[5] Hoff, Syd. *Patty's Pet.* Katonah, N.Y.: Young Readers Press, Inc., 1955.

---

[6] Sarita Bhatnager.

---

[7] Wyler, Rose. *The Riddle Kingdom.* N.Y.: Scholastic Book Services, 1967.

---

[8] Laurie Meunier.

---

[9] Clark, Mae Knight. *Olly's Alligator.* N.Y.: The Macmillan Co., 1968.

---

[10] Tommy Janelle.

---

[11] Kessler, Leonard P. *Kick, Pass and Run.* N.Y.: Scholastic Book Services, 1969.

---

[12] David Morin.

Third Graders

I made a map to show where Columbus sailed.[13] I told about where he went and what he did.[14]

We made a book of main characters. We made pictures and told about them. My main character[15] is fuzzy and has a big nose. He is brown and runs after a bumble bee. He is big and carries a honey pot with him. He has a red shirt and blue pants. My main character is Father Bear.[16]

I made a heating pipe into a bean stalk[17] (Photo 7-3) and dramatized the story with finger puppets. I had to stand on a ladder.[18]

I like to make posters and write a letter to the author. I read Living Free[19] and made a poster.[20]

I made a picture chart by colors (red, yellow, blue and green) of presents suggested by Mr. Rabbit.[21, 22]

---

[13] Lowitz, L. *The Cruise of Mr. Christopher Columbus.* N.Y.: Scholastic Book Services, 1967.

---

[14] Joan Michaud.

---

[15] Berenstain, J & S. *The Big Honey Hunt.* N.Y.: Random House, 1962.

---

[16] Ronnie Theriault.

---

[17] Jones, Jim (Illus.) *Jack and the Beanstalk.* Katonah, N.Y.: Young Readers Press, Inc., 1969.

---

[18] Louise Gorneau.

---

[19] Adamson, Joy. *Living Free.* N.Y.: Scholastic Book Services, 1961.

---

[20] Cara Coro.

---

[21] Zolotow, Charlotte. *Mr. Rabbit and the Lovely Present.* N.Y.: Scholastic Book Services, 1968.

---

[22] Dorothy Fisher.

Photo 7-2.   **Photo 7-3.**

Fourth Graders

I enjoyed making posters and pictures about Abe Lincoln[23] because there were so many interesting things to draw. The most fun though was reading the book.[24]

I enjoyed reading parts of my book[25] orally to others.[26]

Reading *Follow My Leader*.[27] I loved that book. I am sure anyone would like it. I made a cover and wrote about it inside.[28]

What I liked best was writing to an author. When I wrote to Mary C. Jane[29] and got a letter back I found she lived in Maine. Now I have all the things I've gotten from her taped on a big piece of paper.[30]

---

[23] Cavanah, Frances. *Abe Lincoln Gets His Chance*. N.Y.: Rand McNally & Co., 1959.

---

[24] Steven Alexander.

---

[25] Carson, John F. *The Coach Nobody Liked*. N.Y.: Dell, 1966.

---

[26] Luke Hill.

---

[27] Garfield, James B. *Follow My Leader*. N.Y.: Viking Press, 1957.

---

[28] Cindy Nason.

---

[29] Jane, Mary C. *Ghost Rock Mystery*. Philadelphia: J. B. Lippincott, 1965.

---

[30] Jeffrey Simpson.

I made a mobile of Charlie Brown,[31] Lucy, Snoopy and Linus.[32]

We are dramatizing three scenes from a book we all read.[33] We are doing one on Honus Wagner, Ty Cobb and Babe Ruth.[34]

I liked it when two girls and I made a book of characters. It was about Madeline's Rescue.[35] We drew the pictures and after it was finished we made a cover. We enjoyed it very much.[36]

Four of my friends and I wrote a play about Helen Keller.[37, 38]

Fifth Graders

I made a shadow box with another girl. It was about Heidi[39] and her grandmother spinning on a spinning wheel.[40]

---

[31] Schulz, Charles M. J. *We're Right Behind You, Charlie Brown.* N.Y.: Holt, Rinehart & Winston, Inc., 1967.

---

[32] Denise Pratt.

---

[33] Epstein, Sam and Beryl. *Stories of Champions.* N.Y.: Scholastic Book Services, 1969.

---

[34] Chris Murphy.

---

[35] Bemelmans, Ludwig. *Madeline's Rescue.* N.Y.: Scholastic Book Services, 1953.

---

[36] Nancy Tinker.

---

[37] Graff, Stewart and Polly Anne. *Helen Keller, Toward the Light.* N.Y.: Dell, 1965.

---

[38] Gale Carey.

---

[39] Spyri, Johanna. *Heidi.* N.Y.: Scholastic Book Services, 1969.

---

[40] Colleen Bowdoin.

I made a diorama showing Henry and his paper route.[41] I made it out of pieces of cloth, pipe cleaners and yarn.[42]

I enjoyed making a scene of when Miss Bianca, the mouse, is riding the cat.[43] I brought my mouse to school and made the scene around it.[44]

We took the reading lists of every person in the room. (Photo 7-4). We found the books that took place in the U.S. and found out what they were about. We took oaktag paper and drew the U.S. Then we drew the books in the proper states. We also drew small pictures to symbolize them.[45]

I acted the scene when Dr. Doolittle[46] had to go to Africa to help the monkeys from their sickness.[47]

I enjoyed making a poster advertising my book.[48] I made a drawing of a stove. It had two buttons. One had an *M,* the other had an *O. M* meant mischief and *O* meant ordinary.[49]

---

[41] Cleary, Beverly. *Henry and the Paper Route.* N.Y.: Scholastic Book Services, 1968.

---

[42] Donna Dow.

---

[43] Sharp, Margery. *The Rescuers.* N.Y.: Berkley Publishing Corp., 1959.

---

[44] Judy Hubert.

---

[45] Rebecca Lane.

---

[46] Lofting, Hugh. *Story of Dr. Doolittle.* N.Y.: Dell Publishing Co., 1967.

---

[47] Wesley Frewin.

---

[48] Parker, Richard. *M for Mischief.* N.Y.: Scholastic Book Services, 1966.

---

[49] Paula DuBois.

Photo 7-4.   **Photo 7-5.**

I enjoyed making mobiles. I made one about Mexico.[50] The pictures were a sombrero, a bull, the bull fighter and huraches.[51]

I read a book on Maine[52] this year so decided to do a Maine scrapbook. I have 66 pages in it.[53]

I made some apple dumplings and brought them to the class because the book[54] was all about a girl, her apple tree and how she made apple dumplings.[55]

We did a scene of Buffalo Knife[56] in a sandbox (Photo 7-5). We showed the woods and the people going down the river.[57]

---

[50] Wood, Frances E. *Mexico*. Chicago: Childrens Press, 1964.

---

[51] John Thomas.

---

[52] Carpenter, Allan. *Maine*. Chicago: Childrens Press, 1966.

---

[53] Kathy Boucher.

---

[54] Mason, Miriam E. *The Middle Sister*. N.Y.: Scholastic Book Services, 1963.

---

[55] Nancy Waldron.

---

[56] Steele, William O. *The Buffalo Knife*. N.Y.: Harcourt, Brace & World, 1968.

---

[57] Kenneth Chamberlain and Ronnie Shorty.

Sixth Graders

Sixth graders wrote about many interesting projects—writing a poem after reading a ghost story, making a cabin in a tree like Daniel Boone, getting a home run after practicing like Willie Mays, making a giant book cover and writing a brief summary of the book on the back, making a mobile of all the characters in Stuart Little, putting on a puppet show, writing tall tales, dramatizing stories for younger children, improving bowling scores, drawing pictures of horses, making comic strips, reading exciting parts of books to the class, writing a T. V. show and folding paper.[58]

While reading an article by Dr. Thomas Paine[59] in which he describes a trip to Mars it suddenly struck me, "Why can't Doug Plavin and I write a play on an attempt to go to Mars?" That's how "The Attempt" was born. I think that just writing, planning and acting out the play was just as rewarding as doing anything else, but I think the entire play was acted out with a great amount of talent by the actors.[60]

I enjoy making maps or charts and short reports. On Born Free[61] I made a map of Kenya and put all the rivers, mountains, cities and the Adamson's trip on it.[62]

A group of us put on a play.[63] Several classes saw it and hopefully enjoyed it[64] (Photo 7-6).

---

[58] In the order mentioned: Wayne Morey, Tony Brown, Danny Quirion, David Doucette, Anna Quattrucci, Gayle Halliday, Mike Marchetti, Andy Rosenthal, Betty Elliott, Pat Nichols, Billy Dusty, Brenda Hickam, Lee Pratt and Janet Menz.

---

[59] Paine, Thomas O. "Next Steps in Space," *The National Geographic Magazine,* 136: 793–7, December, 1969.

---

[60] Robert Thompson.

---

[61] Adamson, Joy. *Born Free.* N.Y.: Random House, Inc., 1960.

---

[62] Heather MacPhee.

---

[63] White, E. B. *Charlotte's Web.* N.Y.: Dell Publishing Co., Inc., 1952.

---

[64] Jody Jabar.

Photo 7-6.   **Photo 7-7.**

I enjoyed working with papier-mache. Janice Smith and I made a papier-mache octopus and the boy, Mafatu.[65] I thought this was a very interesting book.[66]

Ellen Breard, Jan DeRosby, Ben Reed and I took the last part of the book[67] and made it into a play which we performed for other groups[68] (Photo 7-7).

If it's a good book[69] and other friends like it, we talk about it. I enjoyed making a sculpture of the bear named Gentle Ben.[70]

---

[65] Sperry, Armstrong. *Call It Courage*. N.Y.: Scholastic Book Services, 1969.

---

[66] Linda Towle.

---

[67] Davidson, Mickie. *Helen Keller's Teacher*. N.Y.: Scholastic Book Services, 1969.

---

[68] Kathy Austin.

---

[69] Morey, Walt. *Gentle Ben*. N.Y.: Scholastic Book Services, 1968.

---

[70] David Varney.

From this book,[71] I got instructions for building a TFB universal booster and built one. This is a lower stage that can go on any rocket.[72]

On the cover of Jason and the Golden Fleece,[73] there is a picture of a mosaic of Jason (Photo 7-8). My friend and I are doing a large replica of it with small pieces of construction paper.[74]

I had read a few books[75] where girls had made their own dresses so I decided that I would like to make myself a dress or jumper. I had to read a few books on sewing so I would know how to do basting and so on. After I read the books, I got the material, buttons and a pattern; then I went to work. I had to read the instructions very carefully[76] (Photo 7-9).

I read a book of poetry[77] which inspired me to write my own and draw pictures to go along.[78]

---

[71] Stine, George H. *The Handbook of Model Rocketry*. Chicago: Follett Publishing Co., 1961.

---

[72] Joe Veilleux.

---

[73] Gunther, John. *Jason and the Golden Fleece*. N.Y.: Scholastic Book Services, 1959.

---

[74] Brian Shiro.

---

[75] Alcott, Louisa May. *Old-Fashioned Girl*. Boston: Little, Brown and Co., 1956.

---

[76] Joan McGowan.

---

[77] Read, Herbert (ed.) *This Way Delight*. N.Y.: Pantheon Books, Inc., 1956.

---

[78] Cheryl Skoczenski.

Photo 7-8.   **Photo 7-9.**

Observations teachers have made

Of the activities that were planned and carried out by pupils, puppet shows led the list at all grade levels (Photo 7-10). Plays were also very popular. Young children either read their parts or spoke extemporaneously. Older pupils rewrote stories or certain incidents. Other activities mentioned by teachers of all grade levels were: dioramas and murals, character sketches, book jackets, poetry, clay models, mobiles and posters. All pupils enjoyed reading orally to classmates or to younger children. Most pupils had projects unique to a particular book or story (Photos 7-11, 7-12 and 7-13).

First graders enjoy reading to each other and painting pictures of favorite parts. They also like to write personal experiences similar to those they read about.

Second graders will read to anyone and everyone. Their favorite part is the whole book! They particularly like to record their stories in groups of three or four and then reread as they listen to the recording. Because each person has a different story they pay attention to the child reading and they help each other when necessary. Their puppet shows, roll movies and TV shows will usually consist of a group taking turns reading while others manipulate the characters (Photo 7-14).

Bridget, Kathy and I made a puppet show called the Big Whistle.[79] I helped hold up the puppets but I did not read a word.

---

[79] Beim, Jerrold. *The Big Whistle.* N.Y.: The Macmillan Co., 1968.

It was Kathy and Bridget idea.[80]

I made a play of When I Grow Up[81] whith my friends. We used a table for a airplane.[82]

Second graders also like to make clay models and draw pictures to illustrate their books (Photo 7-15).

Third graders will rewrite stories for plays and puppet shows. They like to draw maps, explain science experiments and demonstrate magic tricks (Photo 7-16). Painting and drawing pictures, making dioramas and posters also are favorites. They can locate and read to the class the most exciting part of a story or the part they like best. They are also happy just to read silently and write about a favorite character. Some begin to take an interest in the authors and write letters to them.

From fourth grade on, pupils choose more written activities. They encourage other members of the class to read the same book so they can work together on a play or puppet show. Sometimes a teacher notes common interests and suggests group projects. Three fifth grade boys became very enthusiastic about reading[83] when brought together this way. The boys[84] substantiated this.

---

[80] Danette Webber.

---

[81] Ainsworth, Judy. *When I Grow Up*. Chicago: Childrens Press, 1968.

---

[82] Lisa Burgess.

---

[83] They listed eight different books. Four were: Asimov, Isaac. *Satellites in Outer Space*. N.Y.: Random House, 1960. Hess, Fred C. *Man in Space*. N.Y.: Doubleday & Co., Inc., 1964. Sutton, Felix. *The How and Why Wonder Book of the Moon*. N.Y.: Wonder Books, 1963. Wyler, Rose and Gerald Ames. *Exploring Other Worlds*. N.Y.: Golden Press, Inc., 1968.

---

[84] Guy Plante, Donald Veilleux, and David Levesque.

I am working on a space project and I have fun doing it. In matter of fact every body likes space projects.

I have got interested in reading after reading my first book about space. I made a model of the blast off.

I am working on a space project. I like space. I have read five books about space. I made a model of the earth. It's the best project that I have ever done.

Upper grade pupils love to read to kindergarten and first graders. They choose appropriate books and practice (Photos 7-17, 7-18, 7-19 and 7-20).

Photo 7-10.   **Photo 7-11.**

Photo 7-12.   **Photo 7-13.**

Photo 7-14.

Well the first thing
that realy had to be
done was to get rid of
the fleas.
For the dog anyways

Photo 7-15.

Photo 7-16.

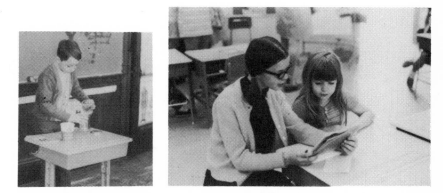

Photo 7-17.

Photo 7-18.

Photo 7-19.

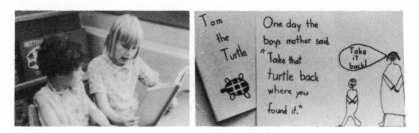

Photo 7-20.   Photo 7-21.

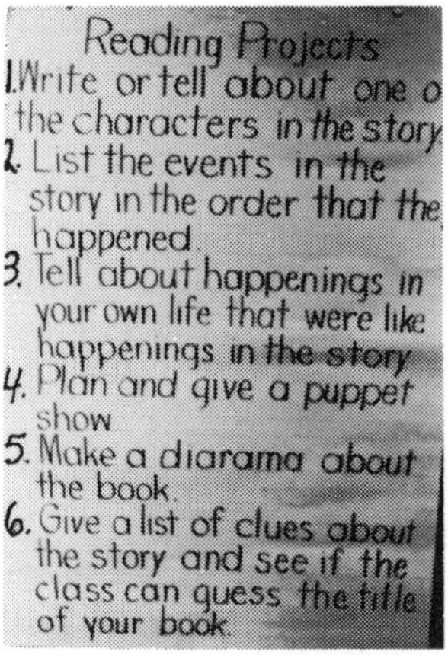

Photo 7-22.

They also write stories and poems for younger children and books for the library. Fourth graders will write and illustrate short stories (Photo 7-21). Sixth graders write book length stories complete with title page, illustrations, chapter titles, index if non-fiction and a suitable cover. They also supply a brief review. The following are samples:

THE MYSTERY OF THE OLD BRIDGE
by Lauren Orloff and Colette Roy

It's about two best friends, Donna Sherwood and Cindy Patterson. Donna's family had a summer camp and Cindy was invited there for the summer.

It all began when Mr. and Mrs. Sherwood and the girls were driving out to the camp and it started raining hard. It rained so hard that they had to stay in a barn for the night. That night Donna was kidnapped!

HERE, POOCH by Anna Quattrucci

Pooch was the Baker's dog, a St. Bernard. If there ever was a dog that got into mischief, this was the dog.

When the dog discovers an old abandoned shack, the excitement begins! It just so happens that a family of cats lives there, and the cats want the shack for themselves.

This is just one of the fun filled adventures Pooch runs into!

WHAT HAPPENS TO POOR SPORTS
by Mark Nale and Joey Joseph

Our story is about poor sports. Poor sports always try to get out of losing whether they know they lose or not. There is a boy in our story named Jeff. Jeff and three other boys start to play baseball but when an other boy starts to hit home runs Jeff gets mad and tries to get revenge.

SHORT STORIES by Brian L.

This book is made up of short stories all written by Yours Truly. Some of the stories are spooky, some are comical, and even others are about war.

One of the stories is about how a little boy named Patrick is turned into a turtle by a witch. You'll rock with laughter at Patrick's episodes.

Another story is about Jim Hallins who is really a werewolf.

Another story is about two soldiers who go to Vietnam. The first day one of the two gets shot. His friend helps him off the battlefield.

NORTH AMERICAN MAMMALS by Heather MacPhee

I wrote my book on animals that live in North America. I gave the general facts on them and after the report I drew pictures of their tracks. As you can guess by the title, I did it only on wild mammals. I am very interested in animals, which inspired me to write this.

Written activities[85] (See Photo 7-22).

1. Write a letter to a friend and tell about the book you are reading.
2. Make a diary as one might have been kept by one of the characters in your book.
3. Make up a crossword or other puzzle about the book.
4. Rewrite the story as a play.
5. Write a description of one of the characters.
6. Write a different ending.
7. Get more information on a topic mentioned in your book.
8. Make a list of unusual words (Photo 7-23).
9. Make a riddle and have others try to guess the name of your book.
10. Write out questions you could ask others who have read the book.
11. Write a book review.
12. Read about the author and write a short biography.
13. Compare two or more books written on the same topic.
14. Add another chapter to the book you are reading.
15. Make a list of other books written by the same author. Read one of them.
16. If it's a story about people and pets, rewrite the story from the point of view of the pet.
17. Find some interesting figures of speech.
18. Write an "autobiography" of one of the characters.
19. Locate your story on a map and then find out something interesting about the place.
20. Find a review of your book and then criticize the review.
21. Make an alphabet book illustrating each letter with a word and picture associated with your book.
22. Compare your book with a movie or TV version.
23. Take an incident from your book and rewrite it as a newspaper might report it.

---

[85] In addition to all teachers mentioned in other parts of this book, the following have contributed ideas directly or indirectly for follow-up activities: Irene Baker, Martha Beach, Nancy Bolio, Lorraine Bowdoin, Phyllis Brown, Ruby Canders, Deanna Cote, Kathleen Covell, Mary Fox, Doris Hesdorfer, Dorothy Holt, Theresa Martz, Martin Rasmussen, Marguerite Sala, Mildred Salsbury, Mary Sawyer, Phyllis Shiro, Patricia Sleamaker, Louise Tracey, and Lois Turcotte.

24. If your book was non-fiction, make a list of questions that were not answered. Look up the answers to your questions and list your sources of information.
25. Make a different title for the story and tell why it would be a good one.
26. Write why you liked or did not like the story.
27. Make a list of some interesting words you didn't know before reading the book.
28. List the important events in the story.
29. Write up the important events in diary form.
30. Rewrite the story as a poem.
31. Rewrite the story for a young child to read.
32. Write a letter to the author.
33. Think of some words that describe the book and tell why you chose them.
34. Rewrite the story or part of the story putting yourself in it.
35. Make a chart of some of your new words making it look like a page from a dictionary.
36. Prepare a statement you could use to sell your book.
37. If there was any part you wish had been left out of the story, tell what it is and why.
38. Tell who else in the class might like the book and why.
39. Tell how one of the characters is like you or someone you know.
40. Choose an incident from the story and tell what you would have done about it.
41. Could the story have happened? Why or why not?
42. Tell what a character did in the story and why you think he did it.
43. What did you learn by reading your book?
44. How do you think you will be different because of the book you read?
45. If your story is about a person who did something, answer the following questions: who? what? where? when? why?
46. Choose one of the characters you particularly liked and tell why you would want that person for a friend.
47. What did you learn about people from reading your book?
48. Write about the funniest or saddest part, the most exciting part or the part you liked best.

49. You are a famous movie star and have been asked to play one of the leading roles in your book. You will not take the part if it differs too greatly from your own personality. Write a letter to the producer telling whether you will take this role, why or why not.

50. Write a note to one of the characters in your book inviting him to dinner. Then write a note you might leave for your mother telling her you have invited someone and tell her something about the person so she will know what kind of meal to prepare.

Dramatic and oral activities

1. Dramatize a scene from your book or put on a puppet or marionette show (Photo 7-24).
2. Present your story in pantomime.
3. Dress up like one of the characters (Photo 7-25).
4. Select a part to read aloud. Tell why you chose it.
5. Read parts of your book to a friend. Then listen to your friend read parts of his book to you.
6. Find a story suitable for a younger child and read it to him.
7. Invite a speaker to talk about a topic related to your book.
8. Using the setting for your book, make believe you had just returned from a trip there and give a travel talk.
9. Tape parts of your book. As you listen to the tape decide how you could have read it better. Try again.
10. Tell about the funniest, saddest or most exciting part.
11. Join a group to discuss books. Each person could tell about a different book, about the same type of book or you could discuss a book all had read.
12. Join with others who have read the same book and present it as a play.
13. After reading a number of books on the same subject, give a short talk while displaying the books.
14. Present your book as a radio book talk.
15. Demonstrate how to make or do something described in your book.
16. Present the story as a TV drama.
17. Have a panel discussion about books related to one topic.

18. Record parts of your story with sound effects and musical background.
19. Organize a quiz show featuring questions about books and authors. (Adapt the format of popular TV shows.)
20. Give an illustrated lecture about your book using a flannel board.

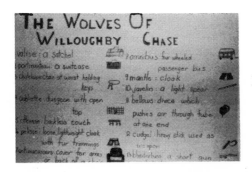

Photo 7-23.

Photo 7-24.

Photo 7-25.

Photo 7-26.

Photo 7-27.

Photo 7-28.   **Photo 7-29.**

Photo 7-30.   **Photo 7-31.**

Photo 7-32.   Photo 7-33.

Photo 7-34.

Photo 7-35.

Arts and crafts activities (Photo 7-26)

1. Draw pictures to illustrate the story (Photo 7-27).
2. Make a book jacket and write something about the book on the inside.
3. Make a poster to advertise your book (Photo 7-28).
4. Make a miniature stage set of a scene from your book (Photo 7-29).
5. Make a model of one of the characters from clay or other material (Photo 7-30).
6. Make a diorama (Photo 7-31).
7. Make a mobile.

8. Make a mural.
9. Organize a book fair or display several books about one topic.
10. Make a map or diagram to illustrate the setting of the story.
11. Make a three dimensional scene in a shoe box (Photo 7-32).
12. Collect objects that would remind one of the story. (The piece of the jig-saw puzzle that was swallowed by Curious George, for example.)
13. Make a film strip to illustrate the book.[86]
14. Make a display of objects or pictures of objects associated with your book.
15. If more pictures were to be added to your book, what would you suggest?
16. If your book is about a family, make a family album by drawing pictures of the family members and writing something about each one (Photo 7-33).
17. Draw or paint a picture to illustrate a scene from your book. Tell why you chose that scene.
18. Make a roller movie or TV show using a box and two sticks or do it in miniature (Photo 7-34).
19. Dress dolls to represent your favorite characters.
20. Copy a good description from your book and then paint or draw a picture to illustrate it.
21. Make some puppets and put on a puppet show.
22. If someone in your book made something, you make one, too.
23. Draw or paint pictures to show how you think the characters looked or how you imagine a scene in the story (Photo 7-35).
24. Make a cube and decorate each side with scenes from your story. You might put the title and author of the book on one side.
25. Make your story into a comic strip.
26. Make a picture dictionary of words associated with your book.
27. Make a travel poster about some place mentioned in your book.
28. Take an incident and try making a rebus story of it.
29. Make a poster showing your favorite books. Make little books out of colored construction paper; put the title on the outside and something about the book on the inside.
30. Make a display of all the books you have read so far this year and take a picture of it.

---

[86] Educational Audio Visual Inc., Pleasantville, N.Y. 10570.

Guiding growth in reading through pupil-teacher conferences

We feel that the success of our individualized reading program is largely due to the pupil-teacher conferences. Pupils share ideas and feelings in a friendly way with a teacher. Getting to know each other as persons builds a mutual respect and understanding. Interviews vary greatly according to the purpose and the personalities involved but some general techniques may be followed.

General techniques

When the language-experience approach is used in kindergarten and first grade, children work alone with the teacher when they dictate their individual experience stories (Photo 8-1). Thus, the conference over a book is a natural step and needs no introduction. When used in situations for the first time explain the procedure to the whole class. Some teachers like to demonstrate with one pupil. This will allay any possible anxiety or curiosity and let you proceed without interruptions. Make the first meeting short in order to reach everyone as quickly as possible. Placing the pupils' names on the board is an efficient way to schedule and serves to keep the class informed. We have tried different ways of scheduling and found the best to be alphabetical order. If you take each person as a book is completed, you will find that some pupils will read short easy books in order to have frequent interviews while others will take difficult books to avoid them. There are many opportunities to talk with pupils between scheduled interviews.

Make the physical arrangements attractive and pleasant and try for some degree of privacy. Have the pupil sit with his back to the class, for example. Some teachers work from a special table while others move about the room. Usually the rest of the class is reading or working on quiet type projects (Photo 8-2).

Photo 8-1.   **Photo 8-2.**

Content

Conferences should be enjoyable discussion sessions, not a question and answer period. This implies that the pupil does as much or more talking than the teacher. A good beginning is to say simply, "What is your book about?" The way this question is answered will reveal much about a pupil's comprehension skills (Photo 8-3). Does he give broad ideas and supporting evidence or does he relate minute details?

> Well it was all about these bears and we[1] made a play out of it and we didn't wear no costumes see because bears don't wear nothing.
> What happened in the story?

---

[1] Second graders. All conferences quoted are from tape recordings.

Sal, she's a little girl, and her mother go picking blueberries and these bears are picking blueberries too and the little girl and the little bear get mixed up. We didn't have no blueberries; we just made believe we were picking them.[2]

Find the part where they get mixed up and read it aloud.[3]

The typical conference begins with some mention of the work done since the last one.[4] If a child has been working on a specific skill, he might be asked to read and react to a short teacher chosen selection or exercise while the teacher is looking over his records. This would be followed by a discussion of a book usually chosen by the pupil. During this discussion the teacher is assessing and teaching comprehension or study skills. Keep a list of these[5] and check according to the type of reading matter. The teacher might then examine the pupil's new word list. If the book is at the correct instructional level, there would be few words. Note the number and the difficulty relative to the pupil's instructional level and immediate needs before deciding what, if any, specific follow-up is necessary. In any event, the pupil should not be required to memorize the words. Older pupils will have few word recognition difficulties but may be concerned with the meaning.

What's the name of the last book you read?

This isn't the last book I read, but this is the book I want to talk about.[6] It's so good that I have two friends reading it and then the three of us are going to do a project together.

How did you happen to choose it?

Well, first the title—I thought it would be about school and then you see this—it won the Newberry Medal so it's supposed to be good.

---

[2] McCloskey, Robert. *Blueberries for Sal.* N.Y.: The Viking Press (Paperback Seafarer Edition), 1948.

---

[3] Ibid, p. 38.

---

[4] See Chapter 11 for samples of Conference Records.

---

[5] See Chapter 10.

---

[6] DeJong, Meindert. *The Wheel on the School.* N.Y.: Harper & Row, 1954.

Was it about school?

In a way. It tells what happened when some children in a one room school try to get storks to come back to their little town. What I liked was that it all started with a story a girl wrote in school for creative writing.

Why do you think it won the Newberry Medal?

It's such a good story you don't want to stop reading it, but I think it's more than that. It gives you a lot to think and wonder about after you finish reading it you keep right on thinking about it. Even in this small town not everyone really knew or cared about everyone else. Then just because one girl wanted to attract storks everyone began helping and soon people began to understand each other. Like, there was Janus, a man with no legs who never went out and no one ever went to see him. Everyone thought he was mean but he really was very nice.

These are my new words. I could pronounce all but the last one, but I wasn't sure of the meanings.

wretchedly
portal
wineballs
morosely
perpendicular
dithering
flotsam and jetsam
soughing

I should have know this one (*perpendicular*) because we have perpendicular lines in math but I couldn't figure what it had to do with a canal. This one (*soughing*) I was sure was printed wrong. I thought it should be *sighing* but it does mean that.

Most teachers have pupils read orally. The pupil may have chosen the selection in advance or the teacher may ask the pupil to read for some specific purpose. The teacher will compare actual performance against the new word list and the pupil's evaluation of the difficulty of the book. Occasionally a pupil will claim he had no problem yet be unable to read satisfactorily. A teacher would diplomatically help him choose easier books. The conference would end with an assignment in skills, if necessary, and a discussion of plans the pupil has for follow-up activities or further reading. The above conference was followed by having a small group share new words.

Each person told what he thought a word meant when presented in isolation. Then the sentence containing the word was read and everyone tried again. Finally the dictionary definitions were read and the group decided on the most appropriate definition. Time for conferences vary, but they will average around ten minutes.

> I'll read this part. It's about the first day Wilbur goes to school and he just sat there and became invisible.[7]
>
> You mean nobody could see him?
>
> Yes because you see he was a ghost. One boy was absent because he was sick and he had left his pencil on his desk so Wilbur picked it up and it looked kind of weird just having a pencil floating around in the air.
>
> Pupil reads.[8] Good. Now from the testing that we did the other day you did very well but there was one thing you were weak on and that was using a dictionary. Today we are going to go down through a dictionary page together and see if you have any questions. Let's begin here. Show me the guide words.

As you get to know your pupils you will concentrate on certain areas omitting those where help is not needed. The amount of time spent with each pupil may vary but the number of conferences should be about equal. Actually immediate help as needed through-

Photo 8-3.   **Photo 8-4.**

---

[7] Spearing, Judith. *Ghosts Who Went to School*. N.Y.: Scholastic Book Services, 1969.

---

[8] Ibid, p. 37.

out the day is the most valuable kind that can be given. Pupils should be encouraged to seek help from fellow pupils before asking a teacher. Giving help may be a greater learning situation than obtaining it (Photo 8-4).

Varying purposes

A conference should not be viewed by pupils as a test, but three times a year this time is used to administer an informal reading inventory.[9] Actually only the first one takes much time because you probably won't know your pupils.

When a class is organizing a special presentation conferences are held to be sure everyone has a plan and knows what to do. The following is an example of a holiday sharing program in which individual and small group conferences played a dominant role.[10]

> Books and magazines from home and from the public, school and class libraries were spread out on a table on Monday for children to examine, discuss and choose. There were two, three and sometimes four copies of the same book so informal groups of pupils reading together resulted. As each individual or group completed the reading they decided on some kind of sharing project to prepare for presentation. *The teacher circulated among the groups*—listened to plans, set up a tape recorder and distributed props and materials as needed for posters and puppets. Rehearsals and preparations commenced as soon as each group or individual could present a plan of action. The program that resulted was shared with a neighboring class (Photo 8-5).

Photo 8-5.

---

[9] See Chapter 9.

[10] Conferences were held by the teacher, Patricia Myers.

Christmas Fun with Books

Watts, Mabel. "Time for the Toys," pp. 36–37, *Jack and Jill,* December, 1969. Two girls took turns reading this picture poem aloud. They also displayed a poster which was a joint effort.

Brown, Michael. *Santa Mouse.* N.Y.: Grosset & Dunlop, 1966. This short story was read orally and a picture exhibited.

Moore, Clement. *The Night Before Christmas.* Racine, Wisconsin: Whitman Publishing Company, 1958. Children dramatized the poem while wearing simple costumes.

Tazewell, Charles. *The Littlest Snowman.* N.Y.: Grosset & Dunlop, 1958. Two girls read a favorite part and showed a poster.

Bedford, Annie North. *Frosty the Snowman.* N.Y.: Golden Press, 1951. Four children gave a puppet show with stick puppets.

Proysen, Alf. *The Town That Forgot It Was Christmas.* St. Louis, Missouri: Webster Publishing Co., 1961. A boy told about the story and read a most exciting part.

Duryeá, Margo. "Plant an Evergreen." *Ranger Rick's Nature Magazine,* Marion, Ohio: National Wildlife Federation, December, 1969. Two boys told about evergreens and showed a chart.

Dempster, Al. *Santa's Toy Shop.* N.Y.: Golden Press, 1950. Children told about the story and showed a book poster.

Hazen, Barbara. *Rudolph the Red-Nosed Reindeer.* N.Y.: Golden Press, 1958. Three children gave a play with costumes and props.

Battaglia, Aurelieus. *The Reindeer Book.* N.Y.: Golden Press, 1965. Two boys read the book and showed favorite pictures.

Brown, Betty H. "Surprise for Santa," *Jack and Jill Christmas Annual.* Philadelphia: Curtis Publishing Co., 1969, pp. 54–55. Two girls in costume acted out the poem.

Dugan, William. *The Christmas Angel Book.* N.Y.: Golden Press, 1965. A girl read a favorite part and showed pictures.

The January *Third Grade News* carried this item:

> Our class made some plays and puppet shows about the stories we read. We made the plays and puppet shows on Wednesday. We made them in our room. We did it because we wanted to have a show. My group made the puppets with sticks. There were other kinds. The people who had the plays wore costumes.

From time to time brief evaluation sessions may follow a group lesson on a specific comprehension or study skill. When this is done each person applies the skill to his own book. The skill may be in some related area as penmanship. An example follows,

> Have the class make a row of letters. Each examines his own and decides which one is best. He circles this one. The class then suggests words that begin with this letter and the teacher writes them on a chart. The teacher then suggests that each person write sentences using as many of the words as possible and even try to add others of their own. As the pupils complete their sentences they have a brief evaluation conference with the teacher. While waiting they engage in other language activities (Photo 8-6).

Since children are writing all the time in a language experience approach this type of lesson periodically makes more sense than formal structured identical practice. Some children like to make individual alphabet books (Photo 8-7).

Photo 8-6.   **Photo 8-7.**

Preparation

Prepare directions for a pupil to follow so that he arrives at the conference with all necessary books and materials in order. They may be placed near the conference list or be stapled inside the pupils' folders. Pupils should be informed of the purpose of the conference in advance so that they come prepared. If they are sharing responsibility for their own learning, they will identify problems and tell what steps they are taking to solve them.

The teacher must review results of the informal reading inventory and notes of previous conferences. Anticipate needs to be ready to offer assistance or suggestions. If you have an organized file of materials, you can supply appropriate aid instantly. An interesting book will often teach a skill better than explanations and artificial worksheets. Keep and continually build a file of suggestions and ideas. An example: If a child tends to over-emphasize the pronunciation of words and ignore illustrations, have him read, *Clifford Gets a Job*.[11] Children love Clifford, a big red dog, and it is impossible to understand the story without consulting the pictures. Another valuable file would be a card for each book with a short statement about the book, a few pertinent questions and a list of follow-up suggestions. The latter might be some that originated with pupils. A lower grade teacher can get a start on such a file with the help of older pupils. We have had sixth graders read the books in a second grade room, state the main idea, provide five questions (and the answers) and then meet with the second graders to try out the questions. This was a valuable experience for all the pupils and although some of the questions had to be revised it gave the second grade teacher a card for nearly every book in her room (Photo 8-8).

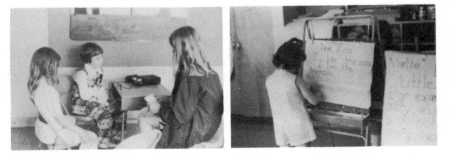

Photo 8-8.   **Photo 8-9.**

Sample conferences

Most first grade conferences are brief and informal. The teacher works with each child at least once a day so a schedule is not necessary.

---

11 Bridwell, Norman. *Clifford Gets a Job*. N.Y.: Scholastic Book Services, 1965.

Do you want me to say all my words?

Let's just pick out some cards.

Grass, flower, finger, problem, photograph, racoon, truck, puck, Rangers, infinity

Infinity? Do you know what that means?

It's like when you're counting and you can just keep right on and on and on and never stop so you say it's infinity.

Are you reading a book?

Yes, but right now I'm writing another story for my book.

Would you like to read your stories instead of a book?

Animals:	Some animals are big. Some animals are small.
Race Cars:	Race cars go fast. There are lots of different races.
Trains:	Trains go fast. Trains are long and big. Some trains are short.
Snakes:	Snakes don't have ears. Snakes are long.
Planes:	Planes can fly. Some planes can land on the snow. There are big jets.
Toys:	Some toys are big. Some toys are little. Toys are fun. Some toys are made of wood.

Children take turns writing their stories on chart paper and on the board and the teacher suggests they read each other's stories (Photo 8-9). One story begins, "I went to a trip to North Carolina," and the teacher stands by as the children work together.

First Child:	I went to a trip . . . .
Second Child:	That's not right. It's on a trip.
First Child:	It says to a trip.
Second Child:	I went . . . I'll have to erase that. There! Now it's *on.*
First Child:	I went on a trip to . . . .
Second Child:	North Carolina. That's where I went.
First Child:	North Carolina.

Since the children are working well together the teacher moves on listening to stories and giving help where needed. When pupils begin reading books they read to each other and discuss their books with the teacher.

Tell me about your story.

I read two.

What were they?

This is about Davy[12] and he goes to school.

Is the story about going to school?

Not all of it. Davy is little then he keeps getting bigger then he gets big then he goes to school.

What were some of the things he could do as he kept getting bigger?

Well he could run. He could jump and he could dress himself.

What did he do to help his mother?

He cleaned with a broom.

And to help his father?

He had a cat in his wheelbarrow.

Is that how he helped his father?

He helped him in the garden.

Did you like that book?

Yes, it was easy.

Here's another one by the same author.[13] You can look at it and you might like to read it. What other book did you read?

This one about a boy he thought he couldn't skate but he could.[14]

Why did he think he couldn't?

He kept falling down. See, here he looks mad.

Look here. Did you do these? (Follow-up on beginning sounds.)

No.

I'll read this part to you. . . . This is a cake. Does cake begin like *can* or like *not?* etc. Then, you can do this other page. (Directions require writing *I can* or *I can not* after looking at pictures.) When you finish show it to me. (See Photo 8-10.)

---

[12] Lenski, Lois. *Big Little Davy*. N.Y.: Henry Z. Walch, Inc., 1956.

---

[13] Lenski, Lois. *Now It's Fall*. N.Y.: Henry Z. Walck, Inc., 1956.

---

[14] Whitman, Tom. *I Can!* N.Y.: American Book Co., 1965.

Photo 8-10.

Conferences vary mainly according to the needs and abilities of the pupils. A group conference is held with second graders in the beginning stages of reading. The teacher had read them *The Tiny Little House*[15] and they dictated a story telling what they would have done with the little house. Following this each pupil wrote and illustrated his own story. The purpose of the conference was to share these stories and then follow up any difficulties. Each child read his story to the group and then they tried reading each other's stories.

> Why is everyone having so much trouble reading yours?
>
> I don't know. I can read it.
>
> I know. It's because he doesn't leave any spaces between the words. Look all along. That's why I can't read it.
>
> How about that? Why didn't you leave spaces.
>
> I don't know.
>
> Let's look at the chart. Who can point to a word? (They each do this.) How many words are in the first sentence?
>
> Eight.
>
> How did you find out so fast?

---

[15] Clymer, Eleanor. *The Tiny Little House.* N.Y.: Atheneum, 1967.

Counted them.

Why was it so easy to count?

The spaces.

Let's look in some books to see if there are spaces between the words. Are the spaces there so that we can count the words?

No, so we can read them better.

And a girl in second grade—

Tell me what your book is about?

It's about the tooth fairy.[16]

What about the tooth fairy?

She comes when you're asleep and she leaves some money and she takes your tooth. She has lots of boxes of teeth and she puts them in jars too and in fish bowls. And she makes jewelry out of them and she makes presents. And she made the path up to her house out of teeth. And she plays games with them with her friends and guess what the prizes are—teeth!

You didn't write down any new words. You knew all the words?

Yes.

Let me ask you some. I'll point and you tell me.

Collected, sort, kitchen, crazy, contest.

Good. Why does the tooth fairy have to work harder than the Easter Bunny and Santa Claus?

Cause they only work one night but she has to work every night because kids are always loosing teeth.

Find that part and read it aloud.

Good. It says she forgets sometimes. Have you ever lost your tooth before you had a chance to put it under your pillow?

No.

Well a girl in a book did and she tried to fool the tooth fairy. You might like to find that book[17] and see what happened.

---

[16] Feagles, Anita. *The Tooth Fairy*. Katonah, N.Y.: Young Readers Press, Inc., 1962.

[17] Hoban, Russell. *What Happened When Jack and Daisey Tried to Fool the Tooth Fairies*. N.Y.: Scholastic Book Services, 1965.

A fourth grade boy—

> My last book was *Miss Pickerell Goes to Mars*[18]
>
> Tell me about it.
>
> There's so much to tell I don't know where to begin.
>
> How did she go?
>
> In a space ship.
>
> What kind of story is it?
>
> It's real funny but it's also about space.
>
> Could you find a funny part and read it?
>
> It's funny all the way through. It's not like all in one place. But I think it's real funny when the captain and Mr. Killian spend a lot of time trying to fix the instruments because they were pointing to 32 instead of 24 and finally the captain said it could only be caused by a magnet and that would be impossible and then Miss Pickerell tells him she has her magnetic hammer with her. I'll find it and read what she says.
>
> So Miss Pickerell's silly remarks make it funny. There's another thing the author does to make it humorous. Do you know what that is?
>
> It's all sort of impossible?
>
> Yes, but what about the cow and the State Fair?
>
> You mean she's always worrying about her cow and her rocks when she should be worrying about getting back? She looks silly too.
>
> Yes. You probably didn't notice the name of the man who illustrated the book (Paul Galdone) but he is a rather famous artist. Sometimes the illustrators are as important as the authors. You can always tell about the stories you read in detail but it's much harder just to give a very brief summary so I want you to try something. Take this page of contents where there are seventeen chapter titles. Try to write just one sentence about each one. Then take the sentences and rewrite them to make one paragraph that will summarize the whole book.
>
> I'm already reading *Miss Pickerell Goes to the Arctic.*[19] I'm going to read all her books.

[18] MacGregor. *Miss Pickerell Goes to Mars.* N.Y.: McGraw-Hill Book Co., 1951.

[19] MacGregor. *Miss Pickerell Goes to the Arctic.* N.Y.: McGraw-Hill Book Co., 1954.

Photo 8-11.

Once the basic comprehension abilities are assessed, and with some pupils this may be done briefly, go deeper or get the pupil's reaction to the book as a whole. Older pupils like to discuss ideas, the author's point of view, the various characters, what the author was trying to convey, etc. They like to test their own ideas on a sympathetic adult. There are times when pupils don't care to discuss certain books with teachers and this is all right, too.

The conference is also a time when teachers give pupils assurance and security and encourage them by pointing out progress. Never feel that a specific skill has to be taught during every conference to make it successful. When you are aware of a pupil's needs you can give help in a variety of situations every day. If others need the same type of help or have read the same book, group conferences are held, but each pupil should have a private time with a teacher to summarize his work, evaluate his progress and plan the next steps. In this way, the pupil is encouraged to share responsibility for his learning (Photo 8-11).

9

Determining instructional needs and goals

Determining instructional levels and needs is one of the most important but difficult phases of an individualized reading program. It is true that some pupils, probably most, will improve in reading without special help just because they continue to read. It is sometimes argued that a poor reader would improve if he would only read more. Unfortunately for him he is usually in a book so difficult that he is too frustrated to learn. In an individualized program, children choose their own books and since no one voluntarily picks a book that frustrates him everyone seems happy. In addition, the reading levels of library books are sometimes so vague that it is difficult to tell just what a child's progress is. Some teachers consider this to be the greatest advantage of the individualized program and are, therefore, reluctant to give even an informal reading inventory.

A point of view

The position taken in this book is that greater progress will be made when the teacher *and* the child know the instructional needs and goals and that these can best be determined with an informal reading inventory.[1] The teacher should assess the child objectively to find out where he is at the moment, how far he can be expected to go, how fast he can be expected to get there, and what the im-

---

[1] A method of combining an informal with achievement tests was presented in Chapter 5.

mediate steps are to get started. You will discover that being objective is the most difficult part of this job. When a child hesitates or stumbles over a word in a test we can't give him credit with, "But I know he knows that word! I taught him that myself!" When he can't answer a question we can't reword it or give him hints. The purpose of the test is to analyze reading difficulties in order to provide effective help. If you keep this purpose in mind, you should have little trouble finding the independent, instructional, frustration and capacity levels for each child.[2]

Reading levels

Independent—Reading level where a person needs no help. Objective measure: Difficulty with no more than one word in a hundred (99%) and comprehension at least 90%. Subjective measure: Oral reading fluent with no signs of difficulty; silent reading with no lip movement or vocalization, finger pointing or other signs of difficulty.

Instructional—Reading level where child should be taught. Objective measure: Difficulty with no more than one word in twenty (95%) and comprehension at least 75%. Subjective measure: Oral reading fluent with little signs of difficulty except for an occasional word; silent reading with *no* lip movement, vocalization, finger pointing or other signs of difficulty. (Note that difficulties at the instructional level are slight.)

Frustration—Reading level where child exhibits various signs of difficulty. Objective measure: Difficulty with more than one word in ten (90% or less) and comprehension less than 50%. Subjective measure: Oral reading with pauses, word by word, voice high pitched or hardly audible, finger pointing, head movement, body movement, yawning, punctuation ignored and words omitted, added or repeated; silent reading with lip movement or audible whisper, finger pointing, head movement, body movements, yawning, slow reading or refusal to read. (Note that *all* signs of difficulty do not have to be present at the same time.)

Capacity—Highest level where child understands when someone reads to him. Objective measure: Comprehension at least 75% based on inferential questions as well as factual which are worded at the same level as the selection.

---

2 Betts, Emmett A. *Foundations of Reading Instruction.* N.Y.: American Book Company, 1957.

Normally the instructional level will be higher than the independent and lower than the frustration or capacity. You cannot assume that they will be a year apart. Some actual findings in a second grade room are: (Numbers refer to graded reading levels—1–1, first half of first grade; 1–2, second half of first grade, etc.)

	PUPIL A	PUPIL B	PUPIL C
Independent	0	2–1	1–2
Instructional	1–1	3–2	2–1
Frustration	1–2	5	2–2
Capacity	3	5	2–2

Before interpreting these scores, let's see how they were obtained. As indicated above you will need word lists for the word recognition scores and "graded" selections to check comprehension. They are available commercially.[3] Those that accompany basal readers are not only more practical but easy to give as the pupils can read the selections right from the book.[4] Word lists can be made up from the basal readers used in your school but the reading selections should be different for those pupils who have read the basals. One based on the Betts Basic Readers[5] that includes only the word recognition test and the selections for oral reading at sight is included here to help you get started. Also included are directions for administering, scoring and interpreting.

Word recognition test

Directions for Administering

Beginning with the primer (1–1) list, have the pupil pronounce each word allowing no more than three seconds for each. Use a card as a pacer. Continue until about half the words in a list are missed. Record the responses on another paper so as not to dis-

---

[3] See footnote 2 on page 72 of Chapter 5.

---

[4] Allyn and Bacon, Inc., American Book Co., and the Macmillan Co.

---

[5] Betts, Emmett A. and Welch, Carolyn M. *Betts Basic Readers*. N.Y.: American Book Co., 1963.

tract the pupil. Use symbols such as (√) meaning word correct, (√^) word incorrect, (O) word omitted, (?) unintelligible. When an incorrect word is given it is better to note it rather than just mark it incorrect. (come/came meaning pupil said *come* for *came*) See Charts 1, 2, and 3.

WORD RECOGNITION TESTS

Primer (1-1)		First (1-2)		Second (2-1)	
Word	Response	Word	Response	Word	Response
1. this	————	1. garden	————	1. ate	————
2. yes	————	2. blow	————	2. bus	————
3. street	————	3. party	————	3. fisherman	————
4. had	————	4. our	————	4. place	————
5. ran	————	5. nose	————	5. men	————
6. kitten	————	6. night	————	6. filling	————
7. things	————	7. parade	————	7. met	————
8. tell	————	8. gave	————	8. cream	————
9. went	————	9. horse	————	9. sister	————
10. one	————	10. would	————	10. goats	————
11. yellow	————	11. road	————	11. evening	————
12. on	————	12. hold	————	12. foot	————
13. happy	————	13. pigs	————	13. need	————
14. girls	————	14. water	————	14. paper	————
15. pies	————	15. last	————	15. opened	————
16. your	————	16. ready	————	16. supper	————
17. time	————	17. fast	————	17. legs	————
18. trees	————	18. right	————	18. stood	————
19. penny	————	19. woman	————	19. still	————
20. there	————	20. home	————	20. presents	————
		21. bat	————	21. I'll	————
		22. pine	————	22. seeds	————
		23. snow	————	23. can't	————
		24. sled	————	24. hour	————
		25. again	————	25. office	————

Chart 1

Second (2-2)		Third (3-1)		Third (3-2)	
Word	Response	Word	Response	Word	Response
1. afraid	————	1. beyond	————	1. world	————
2. pigeons	————	2. butter	————	2. promise	————
3. side	————	3. trouble	————	3. repair	————
4. good-by	————	4. plan	————	4. spoil	————
5. tracks	————	5. angry	————	5. sheet	————
6. bring	————	6. receive	————	6. fort	————
7. hole	————	7. kettle	————	7. load	————
8. fruit	————	8. robber	————	8. shone	————
9. apron	————	9. means	————	9. nickel	————
10. apartment	————	10. castle	————	10. belonged	————
11. nuts	————	11. danger	————	11. twelve	————
12. trips	————	12. handle	————	12. experiments	————
13. telephone	————	13. roast	————	13. packed	————
14. sure	————	14. coyote	————	14. jeans	————
15. moon	————	15. continue	————	15. approaching	————
16. soft	————	16. kept	————	16. halls	————
17. board	————	17. visit	————	17. music	————
18. empty	————	18. circles	————	18. begged	————
19. band	————	19. different	————	19. entered	————
20. air	————	20. cattle	————	20. smart	————
21. hunt	————	21. shape	————	21. folded	————
22. lay	————	22. price	————	22. whip	————
23. true	————	23. turtles	————	23. alarm	————
24. west	————	24. straw	————	24. gay	————
25. rats	————	25. coast	————	25. brayed	————

Chart 2

Fourth		Fifth		Sixth	
Word	Response	Word	Response	Word	Response
1. floated	————	1. behaved	————	1. immediately	————
2. freedom	————	2. pasture	————	2. relax	————
3. usual	————	3. comparing	————	3. plunged	————
4. strength	————	4. mixture	————	4. defend	————
5. explorers	————	5. knitted	————	5. snarl	————
6. witness	————	6. reward	————	6. twine	————
7. obey	————	7. carpenter	————	7. complained	————
8. glanced	————	8. soup	————	8. objected	————
9. ladder	————	9. height	————	9. helm	————
10. foolishly	————	10. polished	————	10. banner	————
11. licked	————	11. simplest	————	11. poisoned	————
12. courage	————	12. determined	————	12. expedition	————
13. fault	————	13. plenty	————	13. initials	————
14. bicycle	————	14. private	————	14. conquests	————
15. report	————	15. motion	————	15. recite	————
16. salt	————	16. horrible	————	16. original	————
17. business	————	17. chips	————	17. rage	————
18. alphabet	————	18. anger	————	18. measure	————
19. elected	————	19. wreath	————	19. fantastic	————
20. stuck	————	20. eagle	————	20. formal	————
21. iron	————	21. concern	————	21. shrank	————
22. northern	————	22. decorated	————	22. verse	————
23. rocket	————	23. celebrate	————	23. depth	————
24. praised	————	24. bracelets	————	24. parchment	————
25. soil	————	25. respect	————	25. lawyers	————

Chart 3

Scoring

Figure the percentage correct at each level and record on a chart. See Chart 4.

	Word Recognition		Comprehension		
	% Correct	Difficulty	Oral	Silent	Listening
Primer (1-1)	————	————	——	——	——
First (1-2)	————	————	——	——	——
Second (2-1)	————	————	——	——	——
Second (2-2)	————	————	——	——	——
Third (3-1)	————	————	——	——	——
Third (3-2)	————	————	——	——	——
Fourth (4)	————	————	——	——	——
Fifth (5)	————	————	——	——	——
Sixth (6)	————	————	——	——	——

Chart 4

Interpretation

The *highest* grade level where the score is 100% may be considered the level where the pupil will have no word recognition problems. If there were words missed below this level or if there were indications of difficulty such as hesitations, saying a word incorrectly and quickly changing it, responding slowly or asking you to allow more time or a second look, you can be sure there will be problems in spite of the 100%.

The *highest* grade level where the score is 95% or 96% may be considered the instructional level if there were no indications of difficulty below that level. Note that this is a tentative instructional level based only on rapid recognition of words in isolation. Pupils can learn to pronounce words yet not know their meanings. Others may know meanings of isolated words but be unable to get meaning from a sentence or longer selection.

Any level where a pupil's score is less than 90% or 92% would be too difficult for him and thus is called a frustration level. Children intensely interested in a subject may read at a frustration level for short periods of time but this should not be encouraged.

Comprehension test

Directions for Administering

Have the pupil read orally at sight beginning with the selection that corresponds to that level in the word recognition test where recognition was 100% with no signs of difficulty. Ask the questions[6] and record the answers verbatim. If the pupil is unable to read this selection fluently with 100% comprehension, drop to an easier one; otherwise proceed to the silent reading selection. Continue until oral reading is characterized by word recognition errors, poor phrasing or other signs of difficulty or until comprehension drops to about 50%. Then begin reading to the child to check his listening comprehension. Continue reading and asking the questions until his listening comprehension drops below 75%.

While the pupil is reading orally at sight mark a copy of the selection for future reference. (You may find this difficult at first. If so, tape the oral reading and make your notations later.) A suggested marking system is:

In ⟨an⟩ individualized	⟨an⟩	word omitted
reading program,	reading	word repeated
each / child / chooses /	each / child /	word by word reading
his ‖‖own reading	his ‖‖own	long pause
material ∧ Some	material ∧	punctuation ignored
children choose ~~very~~ *every* /	~~very~~ *every*	word substituted
easy books ∧ at first *to read*	∧ *to read*	letter omitted; words added
way /✓ This may indicate	way/✓	said *way* and corrected it
FP ⌇ a need to build ⌇ confidence.	FP ⌇	finger pointing

---

[6] Pupils do not look at the selection while answering the questions.

You may want to make notes during silent reading:

FP —finger pointing
LM—lip movement
AW—audible whisper
HM—head movement

SELECTIONS FOR ORAL READING AT SIGHT

Primer (1–1): UP THE STREET AND DOWN[7]

Oral Reading at Sight: 63 words; pages 128–129

Purpose: Find out who the main person is in this story and what that person is going to do.

> One day Sue went
> for a walk in the park.
> Soon she saw something.
> "A penny!" said Sue.
> "I see a penny."
>
> Then she saw Polly and Freddie.
> "I have a penny," called Sue.
> "I will get candy with it."
>
> "Freddie and I will go
> with you," said Polly.
>
> Away they all went.
> They ran out of the park
> and up the hill.

Comprehension Check:
1. Who is the main person in the story? (*Sue*)
2. What did Sue do? (*went for walk in park*)
3. Why do you think Sue went for a walk in the park? (accept any plausible answer)
4. What did Sue find? (*penny*)
5. What was she going to do with the penny? (*buy candy*)
6. Whom did Sue meet in the park? (*Polly and Freddie or two friends*)
7. Why do you think Polly wanted to go with Sue? (*to be sure to get some candy*)

---

[7] Betts, Emmett A. and Welch, Carolyn M. *Betts Basic Readers.* N.Y.: American Book Co., 1963.

8. How did they leave the park? (*ran*)
9. Why? (accept any plausible answer)
10. What do you think is up the hill? (*store*)

First (1–2): AROUND GREEN HILLS[8]

Oral Reading at Sight: 65 words; page 168

Purpose: Find out what the children are going to do.

> Mrs. White looked at the clock.
> But it was not working.
> The hands of the clock were broken.
>
> "Oh, my, the children must not
> be late for school!" she said.
> "And that old clock is broken again."
>
> Just then the children came in.
> All three of them were laughing.
> Jack had a box in his hands.
>
> "Mother," said Jack.
> "Here is a surprise for you."

Comprehension Check:
1. What did Mrs. White do? (*looked at clock*)
2. What did she notice about the clock? (*broken*)
3. Why did she care that the clock was broken? (*children would be late for school*)
4. How do we know it wasn't the first time that the clock had broken? (use of word *again*)
5. Who came into the room? (*children*)
6. How many children were there? (*three*)
7. Why do you suppose they were laughing? (*pleased because they were going to surprise their mother* or any other reasonable answer based on the selection)
8. What did Jack have? (*box*)
9. What did Jack do with the box? (*handed it to mother*)
10. What was the children's last name? (*White*)

Second (2–1): DOWN SINGING RIVER[9]

Oral Reading at Sight: 97 words; page 172.

Purpose: As you read this selection think about the different ways
ways people work.

---

[8] Ibid.

---

[9] Ibid.

Polly Newman and Mary Turner were busy working at a big table.

"Polly and I are building two boats for the show." Mary told the boys.

Dick laughed, "I didn't know girls could build boats," he said.

"Polly and I are a team," said Mary. "We help each other. We're building a team of boats, too."

"They look like work boats," said Bill.

"They are," said Polly. "See, one is long, and the other is short. Do you think we will get the prize?"

"No," laughed Dick. "You see boats like that every day on the river."

Comprehension Check:

1. What were the girls doing? (*building boats*)
2. Why would girls be building boats? (*school project or to enter a contest*)
3. Why did Dick laugh at them? (*probably thought girls didn't know anything about boats*)
4. How did Polly and Mary work? (*as a team*)
5. What does working as a team mean? (*helping each other*)
6. What kind of boats were they making? (*work boats*)
7. Name some work boats. (*tug, fishing, tanker*)
8. How were their boats different? (*one long and one short*)
9. What question did Polly ask Dick? (*what he thought*)
10. Why didn't Dick think the girls would get a prize? (*because their boats were too common*)

Second (2–2): OVER A CITY BRIDGE[10]

Oral Reading at Sight: 118 words; page 191.

Purpose: Find out what important event was going to take place.

Peter didn't think the third Saturday would ever come, but it came at last. He woke up early and ran to the window.

The sky was blue, and the sun was brighter than he had ever seen it. The day would be hot, a good baseball day.

---

[10] Ibid.

"Let's go out and play catch," Peter said to Betty. He wanted to warm up for the game, in which he would play shortstop.

He wanted Betty to warm up, too, for she would be third baseman on the same team.

Secretly, Peter still thought, "Girls can't play baseball very well."

They played catch until noon. Then they got ready for the big game, which would start early in the afternoon.

Comprehension Check:

1. How do we know that Peter was looking forward to Saturday? (*didn't think it would ever come*)
2. Why did he run to the window? (*check on weather*)
3. What kind of day was it? (*sunny, hot, etc.*)
4. Why was Peter pleased? (*wanted to play baseball*)
5. Why did he want to play catch with Betty? (*so both could practice*)
6. What positions would they play? (*third baseman, shortstop*)
7. How long did they practice? (*all morning*)
8. What did Peter think about girls playing baseball? (*girls wouldn't play as well as boys*)
9. How do we know he didn't tell Betty how he felt? (*secretly*)
10. What did they do after they finished playing catch? (*began getting ready for game*)

Third (3–1): BEYOND TREASURE VALLEY[11]

Oral Reading at Sight: 115 words; page 250.

Purpose: Find out about Peter and his family.

Peter Price and his family used to spend the summer months on Frog Island, a small island lying at the mouth of a harbor. The far side of the island faced the deep blue sea.

For Peter, every day on Frog Island was filled with special fun or real adventure. When he wasn't discovering something new, he was busy with his pets.

---

[11] Ibid.

Mr. Price, his father, was head of a zoo. "A love for animals runs in the family," he said. "I suppose you will work with animals, too, just as I do. But whatever job you pick, there's one thing you must learn. It's wise to keep your eyes wide open and notice things."

Comprehension Check:

1. Where did the Price family spend their summers? (*Frog Island*)
2. Where was Frog Island? (*mouth of harbor*)
3. What's a harbor? (*an inlet or bay—where ships anchor*)
4. How did Peter feel about going to Frog Island? (*pleased*)
5. What new things could you discover on an island? (accept all plausible anwers)
6. What else kept Peter busy? (*caring for pets*)
7. Why would his parents probably encourage him to have pets? (*loved animals*)
8. What was Mr. Price's job? (*head of zoo*)
9. What advice did Mr. Price give Peter? (*be observant*)
10. Find two words in this selection that could have two or more common meanings? (accept any that are correct as there are many—*Price, spend, Frog, lying, mouth, etc.*)

Third (3–2): ALONG FRIENDLY ROADS[12]

Oral Reading at Sight: 135 words; page 253.

Purpose: Find out what the Indian boy did.

Once there was a little Indian boy named Shooter of Birds. He could shoot his bow and arrow better than any other boy on the cliff.

One day his mother made him a coat of birds' feathers. The feathers of the robin, the bluebird, the crow, and other birds went into the making of this coat.

Shooter of Birds put on his coat. He took his bow and arrow and descended the path into the canyon. There he went hunting.

But the sun shone so hotly upon him that Shooter of Birds became angry. He shook his bow and arrow at the sun and cried, "I'll shoot my arrow through you, Old Sun."

---

[12] Ibid.

Then Shooter of Birds became angrier than ever. "I'll catch you in a trap," he cried.

Comprehension Check:

1. What was the name of the Indian boy? (*Shooter of Birds*)
2. What could he do better than any other boy? (*shoot a bow and arrow*)
3. What did his mother make him? (*coat*)
4. What color do you think the coat was? (*black, brown, red, etc.*)
5. Why? (*colors of bird feathers*)
6. What does *descended* mean? (*to go down*)
7. Why did Shooter of Birds go into the canyon? (*to hunt*)
8. What made him angry? (*hot sun*)
9. What was the first thing he said he would do to the sun? (*shoot an arrow through the sun*)
10. Why did he become angrier than ever? (*sun laughed at him*)

Fourth: AMERICAN ADVENTURES[13]

Oral Reading at Sight: 151 words; page 210.

Purpose: Find out about turtles.

If the turtle is not scared but only angry, it can shoot its head out and bite with strong, traplike jaws. There are no teeth on the jaws, but the edges are as hard and sharp as the bill of a bird.

All turtles hatch from eggs. In early summer the mother turtle digs a hole in the soft ground or the sand. When the eggs are laid, she covers them with earth and leaves them for the warm sun to hatch.

As soon as the baby turtles hatch, they hurry toward the water, where they feel safer from their enemies, such as crows, gulls, wildcats, and foxes.

In the water the young turtles feed on insects and small water animals. They grow fast, and each scale of the box-like shell adds a new, wide rim every summer. By counting the rims you can tell how old a growing turtle is.

---

[13] Ibid.

Comprehension Check:

1. What does a turtle do when he is angry? (*snap, bite*)
2. How could he hurt? (*edges of jaw are sharp*)
3. In what way are turtles and birds alike? (*hatch from eggs;* be sure pupil's answer reflects some fact mentioned in the reading)
4. How do turtle's eggs hatch? (*sun*)
5. What can a baby turtle do as soon as he hatches? (*run, swim*)
6. Why does a baby turtle run down to the water? (*food, protection*)
7. Who are his enemies? (*crows, gulls, wildcats, foxes*)
8. What do young turtles eat? (*insects, small water animals*)
9. What are the sections of the turtle's shell called? (*scales*)
10. How can you tell a turtle's age? (*A rim is formed between the old scale and the new growth. Since this happens every year age can be determined by counting the rims.*)

Fifth: ADVENTURES HERE AND THERE[14]

Oral Reading at Sight: 195 words; pages 303–304

Purpose: Many years ago a book was written about a famous journey. Find out about this trip.

Marco Polo was a very bright young man who enjoyed looking at people and at things about him. Now he was seeing a great deal, for he was traveling with his father and uncle from Europe to China.

In that year of 1271, Europeans knew nothing of the Far East. Few men dared to brave the long and dangerous journey. But Marco's father and uncle had already been on one trading trip to China. They had even met the great emperor, Kublai Khan, and it was at the Khan's wish that they were returning to the East.

During the long trip, young Marco kept his eyes wide open so he would not miss the wonders of the lands he passed through. Strange, indeed, were the clothes and the ways of the people. The animals and trees were unlike anything he had ever seen or heard of. The palaces and cities seemed more wonderful than those in dreams.

[14] Ibid.

Because he observed so well, and even kept a written record of what he saw, he was able to write a book about China. Even today, it is a book which reads like the most beautiful fairy tale.

Comprehension Check:
1. Who was the traveling storyteller? (*Marco Polo*)
2. Where was he going? (*China*)
3. About how many years ago did Marco Polo make this trip to China? (*700*)
4. Why was it unusual to have Europeans going to China? (*long hazardous trip to an unknown place or any other reasonable answer*)
5. What does it mean "to brave a journey?" (*to have courage to go where others fear to go* or any other explanation that includes the meaning of brave and journey)
6. What was the main reason why Marco Polo and his father and uncle went? (*they were traders*)
7. Who was Kublai Khan? (*ruler of China,* if answer is emperor ask what an emperor is)
8. What did Marco notice about the people and animals? (*different from ones he knew*)
9. What did Marco do after he returned home? (*wrote a book*)
10. What two things did Marco do that helped him write his book? (*observed well and kept notes*)

Sixth: ADVENTURES NOW AND THEN[15]

Oral Reading at Sight: 203 words; pages 312–313.

Purpose: Find out what kind of story this is.

Everyone who lived on the little island off the coast of Panama feared the moon dragon.

Huddled on the earth floor with the rest of her family, a young Indian girl leaned against the bamboo wall of the house. She could feel the wild beating of her heart. Her fear tonight almost drowned out the memory of what the teacher at the Mission School had said this morning.

"Believe me, what I tell you is the truth. It is the truth, Keri."

---

[15] Ibid.

The teacher had told Keri and the other Indian children many tales of things beyond their own islands. And she always ended with the same words; "What I tell you is the truth."

But did the teacher tell the truth about the moon dragon? All of the savage blood in Keri's body denied it. All the teachings of her tribe denied it. Surely her own people knew best what was happening to the moon this very night.

A horrible dragon was out there in the sky, biting pieces from the moon. Everyone must hide, or be bitten, too. Only the moon children, armed with bows and arrows, were safe on this night.

And Ono, her small brother, was a moon child!

Comprehension Check:

1. What kind of story is this? (*legend, folk*)
2. Where does the story take place? (*island off coast of Panama*)
3. Why was the girl's heart beating wildly? (*fear*)
4. What was she afraid of? (*something would happen to her and to her family*)
5. What was going to happen to the moon that night? (*eclipse*; this is not mentioned in the story but no other answer would be acceptable)
6. What were the teachings of the girl's tribe? (*dragon was the cause and he would bite people too*)
7. Why wouldn't the girl's brother have to hide? (*according to their beliefs he would be a moon child so would not be bitten*)
8. Who had tried to tell her the scientific explanation? (*teacher*)
9. Why would it be hard for the children to believe this explanation? (*difficult to accept new idea especially if it is different from family or tribal explanation*)
10. How do legends like this come about? (*a way to explain things we don't understand*)

Seventh: A WORLD OF EVENTS[16]

Oral Reading at Sight: 161 words; page 543.

Purpose: Read to find out something about the sun.

---

[16] Seventh, eighth, and ninth level selections are from: Bailey, Matilda and Leavell, Ullin W. *Worlds of Literature.* N.Y.: American Book Co., 1963.

Our familiar sun, which is a star like the thousands we are used to seeing on clear nights, provides the valuable energy we call heat. It is this energy, pouring earthward across 93 million miles, that eventually causes the various processes we call weather. A breeze, a tornado, a raindrop—all can be traced to the effect of the sun's energy upon the atmosphere.

How important the sun is! It is the center of our family of planets and satellites—the solar system. It provides the core of life's necessities—food and heat. Yet, most of us take the sun for granted, like the air we breathe.

Our earliest ancestors did not take the sun for granted. They realized the power of this glowing ball in the sky and thought it was a supergod. They watched the sun carefully, with fear when it sank low in the sky in the late fall and with rejoicing when it climbed higher in the spring.

Comprehension Check:

1. Explain how weather can be influenced by a star? (*the sun is a star*)
2. If our sun is a star like other stars, why don't we see it at night? (*earth turns on axis so we are turned away from the sun at night* or any other plausible explanation)
3. What makes our sun so valuable? (*provides energy in form of heat*)
4. How does the sun cause changes in the weather? (*by alternating heating and cooling atmosphere, causing water to evaporate* or any other sensible explanation)
5. What is meant by the solar system? (*our earth and planets that revolve around the sun*)
6. Why do most people take the sun for granted? (*movements are predictable* or any sensible explanation)
7. What did the ancient people think of the sun? (*worshipped it*)
8. What in particular frightened the ancient people in the fall? (*when sun sank low they were afraid it would leave forever*)
9. Why would they feel this way? (*they had no scientific knowledge*)
10. What is the main difference between the ancient people and the scientist in the way they viewed the sun? (*ancients were governed by fear; scientists, knowledge*)

Eighth: A WORLD OF EXPERIENCE[17]

Oral Reading at Sight: 236 words; page 569

Purpose: Find out how a young man's enthusiasm influenced photography.

About 70 years ago a young bookkeeper living in Rochester, New York, had a great desire to take pictures, but it wasn't so simple then. In the first place photography wasn't a hobby—it was a profession. So this young man, George Eastman, took lessons from a local photographer. He not only had to learn how to handle the camera, but also the more complicated business of developing the plates and printing from them.

He couldn't afford a studio and, since he had such a great interest in photography, there was only one thing he could do—take pictures outdoors. But you didn't take just a camera with you in those days; you carried an outfit of which the camera was only a part. Eastman's outfit consisted of a camera the size of a soapbox, a large tripod, a big plate holder, a dark tent, a nitrate bath, and a water container.

Taking pictures was more than a hobby in those days—it was an expedition! The young enthusiast soon discovered that he would either have to give up the hobby or get a horse and wagon to haul the equipment. So he decided to do something about it. He analyzed the equipment and found that the tent and the nitrate bath were necessary only because of the wet plates used in those days. So the first job was to get rid of this type of plate.

Comprehension Check:

1. Who was the young inventor in this story? (*George Eastman*)
2. What type of job did he have? (*bookkeeper*)
3. Why did he have to take lessons to take a picture? (*taking a picture also involved developing and printing*)
4. What is the difference between a hobby and a profession? (*hobby usually a spare time occupation*)
5. Why did he have to take his pictures outdoors? (*couldn't afford special studio*)

---

[17] Ibid.

6. What was difficult about taking pictures outside? (*so much equipment was needed*)
7. What did he have to have with him besides the camera? (*tripod, plate holder, dark tent, nitrate bath and container for water were mentioned*)
8. Why did he need a tent and a nitrate bath? (*because he was using a wet plate*)
9. What was the purpose of the plate? (*took image of picture*)
10. Aside from not having to carry around so much equipment what other advantage would a dry plate have over a wet plate? (*picture could be developed later*)

Ninth: A WORLD TO DISCOVER[18]

Oral Reading at Sight: 232 words; pages 532–533

Purpose: Read to find out what the author's experiences were with the Bushmen of Africa.

The first time we came to the Kalahari, we spent several months looking for Bushmen. It was very hard for us to find them because they are shy of any stranger. If they believe that you are coming, they run away like foxes to hide in the grass until you have gone. Their tiny huts, dome-shaped and made of grass, are also inconspicuous. I once walked right into an empty werf, as their tiny villages are called, and didn't see the little scherms, or huts, hidden in the grass until I noticed a small skin bag dangling in a shadow, which was a doorway. Then I saw the frame of the scherm around it, then the other scherms as well. The werf was abandoned, all the people had slipped away, but I heard two voices whispering in the grass near by. You can tell by these werfs that there are Bushmen in an area, and also by their footprints on narrow trails which they share with the game and which run all through the desert. You may find footprints or you may see a little pile of white ash, the fire that a Bushman kindled where he spent the night. Otherwise, to find them you must depend on luck or on the fact that the Bushmen of an area may have heard something good about you and will not be too afraid.

Comprehension Check:

1. In what kind of place does this story take place? (*desert,*

---
[18] Ibid.

but accept answers like *wild grasslands, jungle-like, primitive*)

2. What was the author doing there? (*studying Bushmen*)
3. Why did it take months to find the Bushmen? (*they are shy so run away, they don't stay in one place very long*)
4. Describe the Bushmen's houses. (Accept any description based on the article)
5. What does *inconspicuous* mean? (*difficult to see*)
6. What did the author to do help you know the meaning of *werf* and *scherms?* (*explained in phrase following the words*)
7. How did the author finally find a hut? (*noticed a skin bag*)
8. What two things did the author see that indicated the presence of Bushmen? (*footprints and ashes*)
9. When might a Bushman not hide from you? (*when he has heard something good about you*)
10. Why would a historian find Bushmen an interesting people? (*primitive people* or *live very differently from modern man*)

Scoring and Interpretation

Add comprehension scores to the chart on page 156. The highest level where the comprehension score is at least 90% and the reading fluent is the independent reading level—if the word recognition score at this level is 100%. There may be a discrepancy in these scores. If so, the independent level would be the lower of the two. Except for record keeping this is of little importance. The important thing is to use the results to provide the child with learning experiences that will lead to an improvement in his reading.

Examine the scores further for the highest grade level where comprehension is at least 75% and word recognition at least 95%. Again if there is a discrepancy between these scores take the lower. Then find the frustration level and capacity level according to the criteria stated above. The further the frustration level is from the instructional level, the greater progress can be expected. The capacity level is an indication of a pupil's potential reading ability. The assumption is that a person should be able to read a selection if he can understand it when someone reads it to him. This may underrate some pupils but few read up to their capacity levels. It is used to judge how much progress a pupil may be expected to make.

Sample analysis of three pupils

The scores and the interpretation of those scores made by the pupils mentioned earlier in this chapter now follow.

Pupil A: Independent 0; Instructional 1–1; Frustration 1–2;
Capacity 3–1

	Word Recognition		Comprehension	
	% Correct	Difficulties	% Correct	Difficulties
1–1	100		100	one repetition
1–2	40	relies on	0	couldn't read
2–1		sight words	100	
2–2			80	
3–1			80	
3–2			50	

Pupil A has no true independent level unless it be pre-primer because he relies entirely on memorized sight words. Further testing is necessary to check on auditory and visual discrimination. Until such factors are considered it is difficult to predict his progress but the scores would indicate slow progress because the frustration level is so close to the instructional. It may also mean that he doesn't really understand what reading is. Since he understands third grade level selections when they are read to him, there is no reason to doubt his ability.

Pupil B: Independent 2–1; Instructional 3–2; Frustration 5;
Capacity 5

	Word Recognition		Comprehension	
	% Correct	Difficulties	% Correct	Difficulties
1–1	100			
1–2	100		100	
2–1	100		100	
2–2	100		80	
3–1	100		90	
3–2	96		80	
4	92		70	
5	84	accent	45	80
6	72	slow		25

Pupil B is well above average in ability understanding selections written for pupils in 5th grade which is three years above his grade placement. The few word recognition errors at 3–1, 3–2, and 4th level were careless. At 5 and 6, he didn't know the meaning of the words he mispronounced. A better understanding of syllabication would help him recognize words more quickly. Comprehension difficulties reflect lack of experience. He should be encouraged to broaden his reading interests. Progress will be fast.

Pupil C: Independent 1–2; Instructional 2–1; Frustration 2–2; Capacity 2–2

	Word Recognition		Comprehension		
	% Correct	Difficulties	% Correct		Difficulties
			Oral	Silent	
1–1	100		100	100	voice high
1–2	100	very slow	100	90	long pauses
2–1	100	3 corrections	100	80	repetitions
2–2	80	slow	70		
3–1			40		

This is an example of the importance of considering subjective measures to determine the levels. The scores alone would indicate an independent level of 2–1 but this pupil was very insecure. The higher comprehension scores after oral reading as compared to silent reading probably reflect his greater attention to the task when he felt he was being supervised. The lower scores are more realistic. This pupil needs more reading experiences with easy material and stories he writes himself to give him confidence.

Photo 9-1.

Photo 9-2.

Thus it is possible to establish an independent, instructional, frustration and capacity level for each pupil. The independent and instructional levels may be used as a guide when necessary to help a pupil choose a book or decide which book club to join. Having this information for the whole class would enable you to choose appropriate books and magazines for the room library. The instructional level is the one you will note and perhaps report to parents as the starting point for the year's work. When informal inventories are used regularly throughout the year even small increments of growth can be noted. The frustration level is important in relation to the instructional level as a means of predicting how fast a pupil will progress. When the two are far apart, learning should proceed rapidly. The capacity level will indicate how much progress a pupil can be expected to make. The higher the capacity level the further a pupil can be expected to go. In addition, you will find out what types of word recognition and comprehension problems a pupil has.

None of these levels remain static and they must be interpreted with caution. Their major purpose is not to label or categorize pupils but to help you provide suitable learning activities. How this is done is the subject of the next chapter. (See Photos 9-1 and 9-2).

10

Using reading skills as diagnostic tools

By now it is apparent that an individualized program encourages diagnostic teaching. Pupils are helped in a way that seems most suitable to their mode of learning. Learning to read by a language experience approach is a meaningful extension of oral language development. The teacher's job is to provide appropriate learning experiences. Most pupils learn with seemingly little effort but no matter how well a pupil reads the question of what skills he knows always arises. Perhaps this is not a pertinent question. A list of skills can be valuable as a means of determining what help to give a pupil who is having difficulty. The purpose of this chapter is to show how lists of reading skills can be used as diagnostic tools. They will be arbitrarily classified as word analysis, comprehension and study skills, and oral reading skills.

Word analysis

Basal readers are built on the assumption that there is a clearly defined sequence of skills. These are introduced one at a time and the vocabulary is restricted to words that "follow the rule." Exceptions to the rule are taught as sight words and usually ignored. Research studies are carried out with pupils who have been taught with such basals and under these circumstances it's not surprising when the results "prove" that a definite sequence does in fact exist. The assumption becomes self-perpetuating. One of the interesting

174

aspects of observing children in a language experience approach is the ease with which they learn a wide variety of words. The crucial factor is not their construction but their high personal significance. In fact, when children are given a choice they seldom ask for the little high frequency words that make up so many primary word lists. In spite of this, they learn to read. Why? Probably because of their active involvement in applying all kinds of word analysis skills to material of personal interest. Perhaps also because words are never learned in isolation. This is true even in the early experience charts when a teacher may do the actual writing. A child may paint a picture and after telling about it, ask the teacher to write one word. That one word would have an oral verbal context as well as a picture association. For example, "This is a picture of a squirrel in a tree. Write *squirrel* for me down here." A boy[1] asks a teacher to write *squirrels* on a card for him. Then he writes:

> Fall is interesting.
> The birds fly south.
> Squirrels gather nuts.
> We put pumpkins out.

A girl[2] asks for pretty, pumpkins and clouds as she writes:

> Fall is pretty.
> The leaves are red and yellow.
> Pumpkins are orange.
> The sky is blue.
> And the clouds are wiet.

Another reason they learn is because we emphasize what is correct not what is wrong. Why spoil the joy of such a well composed story by focusing attention on *wiet?* Her spelling makes as much sense, more to her, than the correct spelling. There will be many opportunities to read this word and write it again.

Reading, writing, spelling, listening and speaking are interrelated. Compare the following uncorrected selections written by beginning readers.

---

[1] Robert Levine

[2] Karen Dusty

This is a moonman. It is a martian. Martians are bad. Thay are not real. Martians are stupeid. I came on a spaceship. Spaceships are big.

My creecher is floating in space. My moonman has three ies.

The moonman is walking on the moon. He lives there.

I see a moonmen. He is skinny. He has heavy close.

I see a two headed monster. He is big, big, big, big, big, big.

I saw a monster. He is ten feet tall. He is green. He has 6 arms. He eats green cheese.

A FUN DOG

This is a fun dog.
We have fun with him.
He is a nice dog.
One day we went for a ride.
And the dog was loney.
The dog wilk down the street.
He was very sad because he had know one to play with.
So he ran away.
He saw a boy and the boy play with him.
And little boy Ike him so he brouth him home.
And thay play evry day and thay liv happy.

MY DUCK

this is my little duck Peeking from his little bed I made him. he is a very nice Duck his name is Bubbles. He lives in a very large cage my father made for him. He has a big tub of water this is all about my little duck Bubbles.

Sentence construction, choice of words and attempts at spelling are similar. The space stories were written by first graders in a language experience approach.[3] The story about the dog was written by a twelve year old boy in sixth grade and the one about the duck, a thirteen year old boy in seventh grade. The first graders have had no formal reading, writing, spelling or English while the older pupils have been exposed to five and six years of formal work.

The following are uncorrected selections written by pupils of

---

[3] The three teachers were: Pam Callahan, George Watts and Gary Lisherness.

different grades[4] but all having an instructional reading level of third based on an informal reading inventory.

Grade 1

I went to Quebec last summer. We saw a church and the name of it was St. Anne. We went inside the church. It was very pretty. We went to the zoo. The zoo was very big. It was nice. We went to two museum. One was in St. Anne. The other one was big. I saw the changing of the guard. The guards had a goat. My father took a picture of the goat.

Grade 2

Today is Torrie's birthday. She is nine years old. I can't wait for my birthday. It is June forth and I will be eight. I have to take a little girl to school. She is staying over to my nabors house for ten days. She is in Kindergarten. Her name is Beth. I don't know where she lives. I said I would play with her. Yesterday it was raining. I was just coming home from getting my father. She said, "Hay Kid can you come over?" I said I had to eat.

Grade 3

We made some clay dragons. One has dots all over it. I made a dragon with big bumps on his back. It looks like this (picture). David made one with a flat tummy.

Grade 6

I am like a car on a race track at the Datona 500. I've got speed, I got power, and I got accuracy. I get were I'm going, and I know where I'm going. I stop for fuel just like a car. I also stop for repairs sometimes because I get into accidents too. I pass everyone in sight. I take three pitstops just like in the Datona only mine are breakfast, lunch, and supper.

Grade 7

My uncle was a sailor. He traveled all around the world. He had a lot of fun with the girl. When he was in Hawaii he went surfing and got sun stroke. When he was in Maine he got in a lot of truble and went to jail. That is about the sailor.

---

[4] The five teachers were: Beulah Churchill, Lucille Winters, Patricia Myers, Linda MacLeod and Gary Lisherness.

You might ask yourself some questions.

1. How could a first grader spell so many words correctly without having had formal spelling lessons? (See Photo 10-1.)
2. Why do sixth and seventh graders make errors in spelling when one has had five years of formal spelling and the other six years?
3. Why do first and second graders who have not had English and drill on grammar write just as well as some sixth and seventh graders who have?

Written language for so many pupils has become memorization —memorization of word forms, memorization of phonic and grammar rules, memorization of spelling. Too often the approach is to present one skill after another followed by isolated practice or drill. The teaching of rules and generalizations persists even though they are of doubtful value. Some pupils fail to learn the rules correctly so could not possibly apply them while others give the proper response when discussing them but cannot apply their knowledge to reading or spelling. It would appear that most primary pupils and other beginning readers make no attempt to use vowel rules when trying to figure out unknown words. A more pertinent reason for not teaching such generalizations is that there are too many exceptions. Consider a commonly taught rule, "When there are two vowels side by side, the long sound of the first is heard and the second is usually silent." In that statement there are three words with vowels side by side and not one follows the rule. If you consider that *w* is sometimes a vowel, a child might find five words none of which follow the rule. Of the 220 Dolch words,[5] 35 have two of the usual five vowels side by side and only 10 follow the rule. A popular book with beginning readers is *Blueberries for Sal.*[6] The following are all the words on the first page with two vowels side by side: blueberry, blueberries, brought, berries, our, said, food. None follow the 'rule. Another commonly taught rule is, "When a word has two vowels, one of which is final *e*, the first vowel is usually long and the *e* is silent." The only word that ends in

---

[5] The Garrard Press, 1942.

---

[6] McCloskey, Robert. *Blueberries for Sal.* N.Y.: The Viking Press, Paperback Seafarer Edition, 1948.

e in that statement does not follow the rule. This particular rule is usually presented at the 2–1 reading level. Up to that point, pupils will have learned to recognize such words as: come, have, done, give, gone, care, are, one, where, there, were, some and more. These generalizations have so many exceptions that they are useless.[7] A pupil would have greater success if he were versatile in trying various vowel sounds until the word made sense.

The approach suggested here is to provide the pupil with experiences in listening, writing and reading and to note his progress. As long as progress is steady the teacher would not interfere, but as soon as the pupil has difficulties he cannot solve on his own the teacher provides help. The idea is not to drill on isolated skills but to determine the type of errors that are made and work on the cause of the error. With some children, help may be as simple as making them aware of the problem. A boy who was having difficulty with initial consonant blends solved his own problem by reading *Fox in Sox*.[8]

Using the test of isolated words make a list of all the word recognition errors through the frustration level. After each error note the type—configuration, initial consonant blend, medial vowel, etc. Examine the errors to find a pattern. You may find, for example, that a pupil consistently makes errors in word endings. Another pupil may always get the initial consonant letter correct but guess at the rest of the word. Check the selections to see if the same errors are made in context. It is helpful to ask pupils to explain their method of attack when they make an error. A boy made the following errors: dream/cream, food/foot, your/hour, tricks/tracks, television/telephone, better/butter, anger/danger, continent/continue, castle/cattle. When he was asked, after the testing had been completed, why he said *dream* for *cream* when the word started with a *c* he said that he hadn't even seen the *c*. He gave the same type of response for his other errors. It finally developed that he responded to any known part of a word. If the first part reminded him of a known word, he looked no further. This information could

---

[7] For an excellent summary see, Theodore Clymer, "'The Utility of Phonic Generalizations in the Primary Grades," *The Reading Teacher*. Vol. 16, No. 4, January 1963, 252–258.

---

[8] Seuss, Dr. *Fox in Sox*. N.Y.: Random House, Inc., 1965.

not have been obtained just by studying his errors. Word recognition problems are often related to word meaning. A girl looked at *realm* and said there was an error in printing for *real* didn't have an *m* on it. When words missed are pronounced to older pupils they usually cannot supply a meaning.

After determining the difficulty, choose one or two skills and provide specific help for the pupil. There may be a sequence of skill development but you cannot assume that all children learn according to it. The following list should be used as a guide in assessing a pupil's strengths and weaknesses. It may also assist you in deciding which skills should be emphasized first. If, for example, a child's pattern of errors includes initial and final consonant blends, the sequence would suggest that you concentrate on the initial blends first. When deciding on what help is necessary be sure to note the words the child does know. These suggestions are not to be interpreted to mean isolated drill. Most pupils will progress through wide reading and daily writing.

A list of word analysis skills

The best way to understand the place of word recognition skills as they pertain to reading is to join with other teachers of grades K–8 and try to decide which skills should be taught. Then strive for agreement on a sequence from the easiest to the most difficult. The following is a composite of lists made by some twenty groups of teachers. Use it as an aid in deciding the type of help a pupil needs. Note that some so-called word recognition skills are more appropriate to spelling than to reading.

Has language understandings such as:[9]

> We use words when we talk.
> The words we say can be written.
> Our names are words.
> Words are made up of letters of the alphabet.
> Letters are used over and over again in making words.
> Each letter has a name.
> Letters also have sounds.

---

[9] For more language understandings see Roach Van Allen and Dorris M. Lee, *Learning to Read Through Experiences.* N.Y.: Appleton-Century Crofts, 1963.

What we say can be written.
The name of a story usually tells what the story is about.
The same story can be told using different words.

Uses context, picture, configuration and language rhythm clues to recognize words.

Auditory followed by visual discrimination of:

Rhymes
Initial Consonants
Final Consonants
Medial Consonants
Initial Consonant Blends (Clusters)
Final Consonant Blends
Three Letter Initial Consonant Blends
Consonant Digraphs
Long and Short Vowel Sounds
Effect of *r, w,* and *l* on sound of vowel
Vowel Digraphs
Vowel Diphthongs

Recognizes then uses: (See Photo 10-2).

Various Punctuation
Compound Words (Solid and Hyphenated)
Plurals
Abbreviations
Contractions
Possessives
Inflectional Forms
Prefixes
Suffixes
Root Words

Checking on comprehension

Obviously the ultimate aim of reading is understanding. There are some classroom practices that work against this and you need to be aware of these if you are to give effective help. In some classrooms, there is an inordinate amount of time spent on word analysis and working with isolated words. This is particularly true in rooms where most pupils are working at frustration levels. Pupils become so intent on pronouncing words that they fail to consider the meaning of the selection. They may read word by word or they may read fluently but have no idea what they read. When asked about a story he just read one boy said, "I don't know. I was so busy reading I didn't listen to the story."

Whether or not questions are asked, the kinds of questions asked and the answer accepted will have a direct effect on the type of thinking pupils will do. Ask insignificant factual questions and pupils will read to remember insignificant factual details. Some teachers not only do this but they unwittingly answer the questions themselves by intonation of the voice and by leading the pupils on. Consider this dialogue: "Who spoke first, Sue or Father?" "Sue." "Sue?" "Oh, I mean Father." "That's right. It was Father." Pupils who are not challenged to think as they read have difficulty answering inferential questions. After reading that a little Indian's coat

Photo 10-1.

Photo 10-2.

Photo 10-3.

was made of feathers of the robin, bluebird and crow, one such child said it didn't tell what color the coat was.

Another unfortunate practice is to accept verbatim answers or more information than is called for in the question. For example[10] in answering the question, "How could a turtle hurt you?" a child said almost verbatim, "Well if he's not scared but just angry he can shoot his head out and bite with strong, trap-like jaws and he doesn't have teeth but his mouth is sharp like a bird's bill." The answer is contained in that statement but without further questioning you couldn't be sure the pupil really knows. Further questioning revealed that he did not know. "So how could a turtle hurt you?" "I don't know because he doesn't have teeth but maybe he could because you know my canary sometimes pecks my hand and I can feel it but it doesn't really hurt so I don't think a turtle can hurt you."

The above are examples of ineffective teaching procedures that cause comprehension problems. Not all are due to poor classroom practices. The following are other common causes related to the behavior of the pupils.

1. Word recognition errors
2. Inattention to punctuation
3. Incorrect phrasing (i.e. not reading in thought units)
4. Memorizing words because of failure to understand what reading is
5. Too much reading of material that makes little sense
6. Lack of interest in the story
7. Reading without a purpose or with an inappropriate purpose
8. Failure to adjust rate of reading to purpose
9. Not knowing the meaning of a word or having an incorrect one
10. Choosing an inappropriate meaning for a word
11. Being unaware of specialized vocabulary or special meanings for common words
12. Lack of knowledge of special symbols and abbreviations
13. Inability to recognize and understand idioms

---

[10] All examples are from questions asked about selections from the Betts Basic Readers, American Book Co. See Chapter 9.

14. Becoming lost in details or giving equal weight to all facts
15. Ignoring visual aids—pictures, maps, charts, etc.
16. Misunderstanding pronouns and other referants
17. Lack of understanding of organization of paragraphs
18. Poor memory
19. Inability to concentrate
20. Language disorders
21. Insufficient background of experience for the selection
22. Inadequate concept development
23. Inability to visualize story settings and actions
24. Emotional problems
25. Refusal to consider the author's point of view

An informal reading inventory may be used to assess comprehension. Generally there are three types of questions—factual, vocabulary and inferential. Begin by classifying the questions missed. Sometimes it will be obvious to you. At other times, you may want to ask the pupil to explain his answers. After reading a selection on turtles[11] which contains this sentence, "When the eggs are laid, she covers them with earth and leaves them for the warm sun to hatch,"[12] a pupil said that turtles cover their eggs with leaves. That is an obvious error. Another pupil reading the same selection could not explain satisfactorily how to determine a turtle's age. When asked why she was having difficulty she said that she really didn't know what a scale was. Another girl gave an accurate explanation but had no idea where to look to determine how old the turtle in the classroom was. There are levels of comprehension and within these levels, there are varying depths of understanding. Some pupils have difficulty because they don't realize the relationship between spoken and written language. They don't attempt to "hear" the intonation of the writer or visualize his manner of speaking. Words don't bring them a mental picture. It is possible to explore these areas of comprehension by following up a pupil's answers and by more open or undirected questioning. For example, instead of

---

[11] From Book 4, *American Adventures,* of the Betts-Welch Basic Readers, American Book Co. (See Informal Reading Inventory in Chapter 9.)

---

[12] Ibid, p. 210.

asking the questions in order begin with, "What was that selection about?" or "Tell me all you can remember." Then ask questions over the parts left out. Such an approach reveals a pupil's method of learning. Some will give an almost verbatim account while others will state the main idea. Others will remember nothing (unaided recall) but still be able to answer the questions (aided recall).

The relatively simple comprehension skills can be checked with an informal reading inventory, but working toward higher levels of understanding is a continuous process. Have a list of objectives to guide you in helping each pupil during conferences. The following are suggestive.

Comprehension "skills"

General
1. Understands the main idea
2. Notes details
3. Follows a sequence of events
4. Perceives relationships such as cause and effect
5. Recognizes important and unimportant facts
6. Makes inferences
7. Predicts outcomes
8. Makes judgments
9. Evaluates a story
10. Recognizes fact and opinion
11. Interprets the emotional tone
12. Appreciates the author's style
13. Relates to own experience
14. Classifies and summarizes ideas
15. Compares two or more selections

Factual Material
1. Relates facts to own purpose
2. Selects pertinent facts (uses facts to answer questions or prove a point)

3. Follows directions
4. Draws conclusions from facts
5. Infers from related facts
6. Distinguishes facts from opinions
7. Notes significant details
8. Generalizes
9. Compares facts from more than one source
10. Comes to logical conclusions
11. Makes critical judgments
12. Organizes facts to remember
13. Acquires new vocabulary
14. Extends concepts
15. Considers author's background and motives

Literary Material

1. Identifies literary types
2. Understands mood
3. Identifies main and supportive characters
4. Visualizes characters, setting, action
5. Recognizes character traits
6. Follows sequence of events
7. Predicts outcome
8. Summarizes
9. Interprets author's meaning
10. Appreciates author's choice of words, use of humor, descriptive passages, etc.
11. Judges literary quality
12. Enjoys different types of literature—dramatics, poetry, etc.
13. Desires to own books
14. Shares books with others—dramatizes, rewrites in different form as for puppets, illustrates, etc.
15. Makes personal use of ideas gained from reading or experiences something of personal siginficance

Reading is not a separate subject in an individualized reading program. Some pupils get interested in studying specific topics that may or may not be related to the immediate curriculum. Use this opportunity to check on and teach study skills. All pupils have need of study kills though, as transition from literary type material to science and social studies is not always easy. Be aware of these different skills so that you can help your pupils become versatile readers.

STUDY-TYPE READING SKILLS

1. Skims for dates, places, key words; to locate information quickly; to select pertinent material; to identify main thought; to locate specific facts or details; to review
2. Varies rate of reading according to purpose
3. Follows written directions
4. Rereads when necessary
5. Interprets pictures, diagrams, maps, charts, tables, graphs
6. Notes special and technical vocabulary and symbols
7. Reads for immediate and delayed recall, to find answers to specific questions, to organize ideas, to take notes
8. Deals with various types of test questions—true-false, multiple choice, matching, completion, identification, essay
9. Organizes material—outline, summary, written and oral reports
10. Evaluates material—relevancy, author's qualifications and purpose
11. Knows parts of books—title, copyright, table of contents, index, glossary, appendix, footnotes
12. Appreciates organization of books—chapters, sections, headings, sub-headings
13. Locates and uses sources of information—encyclopedias, yearbooks, almanacs, atlases, globes, special dictionaries
14. Uses library effectively—card catalogue, arrangement of books on shelves and in stacks, Readers' Guide, special indices
15. Makes use of study skills in school subjects and in independent research

DICTIONARY SKILLS

1. Alphabetizes

2. Uses entry words
3. Uses guide words
4. Chooses appropriate definition
5. Uses pronunciation key
6. Understands diacritical marks
7. Understands phonetic respellings
8. Understands syllabication
9. Understands primary and secondary accents
10. Understands abbreviations for parts of speech
11. Understands derivitives of words
12. Understands preferred pronunciations
13. Familiar with other uses of dictionary (common names, geographical names, table of weights and measures, etc.)
14. Familiar with different types of dictionaries and their uses
15. Makes use of dictionary as needed

Oral reading skills

There are many opportunities for meaningful oral reading in an individualized program. From the very first language experience charts, children read to the teacher and to each other. Beginning readers read to classmates and older pupils read to younger. Pupils share their books in small groups and in whole class activities. Because everyone has something different to offer there is no problem getting others to listen. Good oral reading requires (Photo 10-3):

1. Fluency
2. Conversational tone
3. Correct interpretation of punctuation
4. Good enunciation
5. Pleasing voice
6. Sufficient volume for size of room
7. Ability to set and maintain mood of selection
8. Awareness of audience reaction
9. Ability to choose pertinent selections
10. Desire to share reading with others

11

Recording and evaluating progress in reading

Record keeping

When every pupil is reading a different book, working in a different skills area and making a variety of projects, it becomes difficult to remember what anyone has accomplished (Photo 11-1). In order to have any kind of evaluation, it is necessary to keep records, but they need not be elaborate. In fact, they should be simple enough to give the desired information in the shortest possible time. Since no two teachers will ever agree on what that desired information is you will find many different types of records.

Pupils' records

In individualized reading you keep records, but they are not hard ones to keep. You keep a record of how many pages you read a day and all the books you've read.[1]

Keeping records isn't so much fun but it does give you responsibility.[2]

[1] David Doucette.

[2] Paul Carey.

We must write down and record what we have read. If the teacher asks us to recommend a book, we can use our reading records to see what book we liked.[3]

Beginning readers at every grade level from kindergarten on are expected to keep a list of the books they read. Rule off the inside of a folder for this. The same folder is then used to hold other papers. Kindergarten and first grade children probably will not keep any other records regardless of their reading ability. They will be writing and working with words in a language experience approach. They learn the mechanics of writing very quickly. The teacher makes sure the child watches as she writes his dictated words so that he sees how letters are formed as he learns them. Some pupils begin by copying the teacher's work. A teacher might then use pencil and suggest that the child trace over her writing with crayon. As children become adept at writing they keep more records. At the same time, they will be reading so much more that they will keep their records on separate papers in a folder. These can be stapled together from time to time and stored for future reference. Two other records will be added to the list of books[4] by title and author. One of these will be a daily log to show: date, title of book, author, pages read and sometimes a place to note other work done. The other will be a record of new words. These may be used during the conference to check on the difficulty level of the book and as a basis for discussing word learning techniques or word meaning. Most teachers have their pupils complete a form when they finish a book before they read another book or begin a project (Photo 11-2). This usually calls for a statement or a picture—tell what the story was about or draw something interesting about the story. Children could choose from a variety of forms. Ours are mimeographed in colors.

In some classrooms, particularly those in which reading is considered one of the language arts, pupils would keep no other records for there would be many follow-up activities.

---

[3] Emily Bridgman.

---

[4] The word *book* is used but magazines, newspapers, almanacs, encyclopedias, etc. may be read.

Photo 11-1.   **Photo 11-2.**

Sample Forms That Will Be Helpful

For beginning readers

Name			Pages Read		
Date	Title	Author	Start	Stop	Other Work

Daily Reading Record

Name: _____   Date: _____

Title: _____   Author: _____

New Words	Draw a picture or write something interesting about the story.
_____	
_____	
_____	
_____	
_____	
_____	
_____	
_____	
_____	
_____	
_____	

Name: _____   Date: _____

Title: _____   Author: _____

Draw something about the story.	Write something about the story.

New Words

_____

_____

_____

Name: _____ Date: _____

Title: _____

Author: _____

New Words	This story was
_____	easy hard
_____	just right
_____	

Tell what the story was about. _____

_____

_____

_____

_____

_____

Title: _____ Name: _____

Author: _____ Date: _____

Three questions about the story:

1. _____

2. _____

3. _____

Answers:

1. _____

2. _____

3. _____

For advanced readers

READING RECORD   NAME _____

          TITLE   AUTHOR   COMMENTS

1. _____
2. _____
3. _____
4. _____
5. _____
6. _____
7. _____
8. _____
9. _____
10. _____
11. _____
12. _____
13. _____
14. _____
15. _____
16. _____
17. _____
18. _____
19. _____
20. _____
21. _____
22. _____
23. _____
24. _____
25. _____

Instead of "New Words" pupils might note:

Types of words such as action, descriptive, sensory, color, humorous, mysterious, etc.

Idioms, metaphors, similes, alliteration

Words that convey shades of meaning such as words that mean "to ask," "to go," "big."

REPORT ON BOOK READ

Title _____ Name _____

Author _____ Date _____

What type of book is it? _____

Where did the story take place? _____

When? _____

Who were the main characters? _____

_____

Describe one of the characters _____

_____

_____

Tell one important thing the above character did. _____

_____

_____

Why did he do it? _____

Would you have done it? _____

Why or why not? _____

_____

_____

Title of Book _____ My Name _____

Author _____

What was the book about? _____

_____

_____

Where did the story happen? _____

What was the most interesting part? _____

_____

_____

Tell why someone would like the story. _____

_____

_____

_____

Special records

Other records that might be used from time to time check on reading, comprehension or study skills, or are designed for specific types of books such as biography or poetry. The latter are usually used in the upper grades.

WORKSHEET ASSIGNMENT

Name _____ Week of _____

Monday: Sheets _____ _____ _____   Checked with:

        Score _____ _____ _____

                                    (Friend's Name)

Tuesday: Score _____ _____ _____

        Score _____ _____ _____   _____

WORKSHEET ASSIGNMENT
(Do pages circled)

Name _____

Date	Page	Kind of Worksheet	Possible Score	Pupil's Score	Comments
	1	Following Directions	10		
	2	Sentence Comprehension	15		
	3	Classifying Ideas	8		
	4	Locating Information	12		

DICTIONARY SKILLS

Name _____

Date _____

Title of Book _____

Author and Publisher _____

Sentence Containing Word (Underline the Word)	Page	Dictionary Respelling	Meaning

VOCABULARY LIST

Title of Book _____ Name _____

Author _____

Date	Word	Sentence and Page No.	Definition

REPORT ON BIOGRAPHY

Title _____   Name _____

Author _____   Date _____

Who? _____

When did he live? _____

Where did he live? _____

What made him famous? _____

_____

Tell about his early life. _____

_____

_____

Tell about his later life. _____

_____

_____

What kind of person was he? _____

_____

Tell why you would or would not like to be like him. _____

_____

_____

_____

Teachers' records

There is no need for teachers to duplicate records kept by pupils. Teachers will keep two kinds—one to show reading levels at least three times a year and another to summarize the pupil conference. The first should become part of each pupil's permanent folder because it will indicate the immediate instructional level and needs, how much the pupil may be expected to accomplish and how fast he should accomplish it. It should also include the recommended instructional level for the next school year.

INDIVIDUALIZED READING PROGRAM

Pupil _____ School _____

Teacher _____ Year (Sept. _____ June _____)

	Test _____		Test _____		Test _____	
	Word Recog.	Comp.	Word Recog.	Comp.	Word Recog.	Comp.
PP	——	——	——	——	——	——
1-1	——	——	——	——	——	——
1-2	——	——	——	——	——	——
2-1	——	——	——	——	——	——
2-2	——	——	——	——	——	——
3-1	——	——	——	——	——	——
3-2	——	——	——	——	——	——
4	——	——	——	——	——	——
5	——	——	——	——	——	——
6	——	——	——	——	——	——
7	——	——	——	——	——	——
8	——	——	——	——	——	——
9	——	——	——	——	——	——

Independent ——————————————— ——————————— ———————————

Instructional ——————————————— ——————————— ———————————

Frustration ——————————————— ——————————— ———————————

Capacity ——————————————— ——————————— ———————————

Specific Needs:

Date _____

Date _____

Date _____

Recommended Instructional Level for Sept. _____ _____

                                        year   level

Conference records may vary from a checklist to a summary.
The following is a sample:

CONFERENCE REPORT

Pupil's Name _____ Date _____

Title and author of book discussed _____

_____

Word recognition and meaning _____

_____

Comprehension _____

_____

Oral reading _____

_____

Immediate needs _____

_____

Specific assignments (if any) _____

_____

Project and other work planned _____

_____

Progress noted _____

_____

Other comments _____

_____

Evaluation Techniques

Evaluating pupils

Evaluation is continuous in an individualized program since instruction is based on immediate needs as determined at first by
acquisition of language understandings and later by a reading inventory and the problems that show up in conferences. You will also
have each pupil's independent, instructional, frustration and capacity levels, and from these you will know where the pupil is (in-

structional level), how far he is capable of going (capacity level), how far he can be expected to go within a reasonable length of time (frustration level) and the rate of expected progress (the greater the difference between the frustration and instruction levels, the faster the expected progress). In the middle of the year and again at the end, different reading inventories are administered and the results compared. A change in the immediate needs or in any of the levels denotes progress. It is possible for the instructional level to remain the same while the independent and frustration levels move up. In any event, the second inventory must be interpreted in relation to the first and it is possible when necessary to indicate small increments of growth. Other accomplishments can be noted by referring to the conference records.

Keep dated samples of pupil's work to discuss with them, to use in conferences with parents and to send on to the next teacher. Pupils who had a language experience approach from the beginning of kindergarten wrote for ten minutes every day during second grade on topics of their own choice. In the following samples, observe not only the individual growth but also the range of ability.

Elias   Oct. 28

on friday satdaday and sunday
I was sik.
I did not go out for 3 day.
I went out on Monday.

May 26

Yestuday I wocht the Red Sox game. They won 5 to 3. Carl Yetremsky hit a home run and Tone Cniglearo hit two home runs.

Paul   Oct. 20

Peter and me are gowing
to maek a book
We are gowing to maek
a book allbiyare selvs
and we will have a good time

May 27

Today is gym. I hope we play kickball. I have only kicked one home run.
Last night I lisend to the Red Sox. They lost 5 to 7. They played the Senetors. Tony Coniglearowe tripeled two times and drove in two runs.

Peter   Oct. 20

Today I brought
my frisbee.
It is a red one.

May 27

We are painting our kitchen. It is going to be yellow. We are almost done. All we have to do is

It can fly good.
When it flys
It goes High.

put another coat on one wall. It is pretty. We have to do the doors of our cabinets. Some of them are finished and some of them are not.

Bill   Oct. 20

We are rireing (writing) for ten manit. (minutes)

May 28

I have posnive. (poison ivy) I am going fishing for two days. Did you ever see a dog jog? Monday I will bring a picher of one. I will bring something all so. It is my notebook.

Beth   Oct. 23

Today I have bronies in the audotorium. I was chaed to threrday insted of Fryday becouse thir was tomany.

May 28

Today is fly up. The second year Brownies will fly up. But the first year Brownies will not. They will have duties. Christina's job is to seat the parents. Ronica's job is to greet at the door. My job is to serve the people. I hope we have fun. Do you go to Brownies? Next year I will fly up.

Reporting a child's progress to parents is an individual matter. We try to give everyone the same basic information but explain it in a way a parent will understand. When parents come for their conference they also discuss their child's records and see various projects and work samples displayed around the room. Some parents are well informed by their children who are quite aware of their own problems and progress. The brief outline for the reading section of the parent-teacher conference follows:

READING—Individualized Program

Instructional Level:

Immediate Instructional Needs:

    Word Recognition
    Comprehension
    Other (Oral reading, work habits, creativity, etc.)

Predicted Rate of Progress:

Review of Pupil's Records: (List of books read, record of work done, sample of follow-up activities, etc.)

There are two other areas to consider when evaluating pupils— application of knowledge and quality of work done. We must not stop with the question, how well does he read? It is just as important to ask, does he read? Regardless of a pupil's instructional level we must encourage and expect a seriousness of purpose that results in pride of accomplishment. Individualized reading does not imply lowering of standards, but the goals for each pupil are realistic. A sixth grade girl had this to say in one of her conferences:

> Sometimes when I look at the books some of the other kids are reading I get discouraged but then when I think how low I was when I came into this room I get encouraged all over again. I really have made a big improvement in my reading and besides I can enjoy it now.

Individual and group tests can be given to check on skills. Sometimes a pupil is given a brief test over one specific skill during or just before a conference. At other times, a group or the whole class is given an achievement test. Standardized test results will reflect greater gains than usually expected, but such objective measures are limited. Other factors to consider are (Photo 11-3):

1. The greater number of books read
2. Development of wide interests and enthusiasm for reading
3. Growing ability to evaluate books critically
4. Awareness of authors, illustrators and literary style
5. Evidence of application of reading-study skills in all areas of the curriculum
6. Increase use of dictionaries, encyclopedias and other reference books

Photo 11-3.

7. Improvement in oral reading
8. Growth in vocabulary and interest in words
9. Ability to discuss and react to books and make personal use of reading
10. Assumption of greater responsibility for learning

Selected Bibliography

Allen, Roach Van and Claryce Allen. *Language Experiences in Reading.* Chicago: Encyclopaedia Britannica Press, 1966.

Almy, Millie, Edward Chittenden and Paula Miller. *Young Children's Thinking.* N.Y.: Teachers College, Columbia University, 1966.

Applegate, Mauree. *Freeing Children to Write.* N.Y.: Harper & Row, 1963.

Ashton-Warner, Sylvia. *Teacher.* N.Y.: Simon & Schuster, 1963.

Barbe, Walter B. *Educator's Guide to Personalized Reading Instruction.* Englewood Cliffs, N.J.: Prentice-Hall, 1961.

Betts, Emmett Albert. *Foundations of Reading Instruction.* Chicago: American Book Company, 1957.

Bower, Eli M. *Early Identification of Emotionally Handicapped Children in School.* Springfield, Ill.: Charles C Thomas, 1960.

Bruner, Jerome S. *The Process of Education.* Cambridge, Mass.: Harvard University, 1960.

Bruner, Jerome S., Rose R. Oliver and Patricia M. Greenfield. *Studies in Cognitive Growth.* N.Y.: John Wiley & Sons, 1966.

Burrows, Alvina T., Doris C. Jackson and Dorothy O. Saunders. *They All Want to Write.* N.Y.: Holt, Rinehart and Winston, 1964.

Calder, Clarence and Eleanor M. Antan. *Techniques and Activities to Stimulate Verbal Learning.* N.Y.: The Macmillan Co., 1970.

Cushenbery, D. C. *Reading Improvement in the Elementary School.* West Nyack, N.Y.: Parker Publishing Company, Inc., 1969.

Darrow, Helen and Virgil M. Howes. *Approaches to Individualized Reading.* N.Y.: Appleton-Century-Crofts, 1960.

Dechant, E. *Diagnosis and Remediation of Reading Disability.* West Nyack, N.Y.: Parker Publishing Company, Inc., 1968.

deMille, Richard. *Put Your Mother on the Ceiling.* N.Y.: Walker and Co., 1955.

Doll, Ronald G. *Individualizing Instruction.* Washington, D.C.: Assoc. of Supervision and Curriculum Development, 1964.

Eash, Maurice J. *Reading and Thinking.* N.Y.: Doubleday, 1967.

Fader, Daniel N. and Morton H. Shaevitz. *Hooked on Books.* N.Y.: Berkley Publishing Co., 1966.

Fernald, Grace M. *Remedial Techniques in Basic School Subjects.* N.Y.: McGraw-Hill, 1943.

Flanders, N. A. *Teacher Influence, Pupil Attitudes and Achievement.* Minneapolis: University of Minnesota, 1960.

Flavell, John H. *The Development Psychology of Jean Piaget.* N.Y.: D. Van Nostrand Co., 1963.

Goodlad, John I. and Robert H. Anderson. *The Nongraded Elementary School.* N.Y.: Harcourt, Brace & World, 1963.

Goodlad, John I. et al. *The Changing School Curriculum.* N.Y.: The Fund for the Advancement of Education, 1966.

Hall, MaryAnne. *Teaching Reading as a Language Experience.* Columbus, Ohio: Charles E. Merrill Publishing Co., 1970.

Holt, John. *How Children Fail.* N.Y.: Pitman Publishing Co., 1964.

Holt, John. *How Children Learn.* N.Y.: Pitman Publishing Co., 1967.

Inhelder, Bärbel and Jean Piaget. *The Growth of Logical Thinking.* N.Y.: Basic Books, 1958.

Klausmeier, Herbert J. and C. W. Harris. *Analyses of Concept Learning.* N.Y.: Academic Press, 1966.

Lee, Dorris M. and Roach Van Allen. *Learning to Read Through Experience.* N.Y.: Appleton-Century-Crofts, 1963.

Martin, W. T. and Dan C. Pinck. *Curriculum Improvement and Innovation: A Partnership of Students, School Teachers and Research Scholars.* Cambridge, Mass.: Robert Bentley, Inc., 1966.

Martin, William (ed.) *Sounds of Language Readers.* N.Y.: Holt, Rinehart and Winston, 1966.

Miel, Alice (ed.) *Individualizing Reading Practices.* N.Y.: Teachers College Press, Columbia University, 1958.

Minuchin, Patricia, Barbara Biber, Edna Shapiro and Herbert Zimiles. *The Psychological Impact of School Experience.* N.Y.: Basic Books, 1969.

Moffett, James. *A Student-Centered Language Arts Curriculum, Grades K-13.* Boston: Houghton-Mifflin Co., 1968.

Myers, R. E. and E. Paul Torrance. *The Ideabook Series in Creative Development.* N.Y.: Ginn and Co., 1966.

Parnes, S. J. and H. F. Harding. *A Source Book for Creative Teaching.* N.Y.: Charles Scribner's Sons, 1962.

Petty, Walter T. and Mary Bowen. *Slithery Snakes and Other Aids to Children's Writing.* N.Y.: Appleton-Century-Crofts, 1967.

Piaget, Jean and Bärbel Inhelder. *The Psychology of the Child.* N.Y.: Basic Books, 1969.

Reasoner, Charles F. *Releasing Children to Literature.* N.Y.: Dell Publishing Co., 1968.

Robison, Helen F. and Bernard Spodek. *New Directions in the Kindergarten.* N.Y.: Teachers College, Columbia University, 1965.

Sanders, N. M. *Classroom Questions.* N.Y.: Harper & Row, 1966.

Scott, Louise Binder. *Learning Time with Language Experiences for Young Children.* St. Louis: Webster Division, McGraw-Hill, 1968.

Shaplin, J. T. and H. F. Olds. *Team Teaching.* N.Y.: Harper & Row, 1964.

Smith, James A. *Creative Teaching Series.* Boston: Allyn & Bacon, 1967.

Stahl, D. and P. Anzalone. *Individualized Teaching in the Elementary School.* West Nyack, N.Y.: Parker Publishing Company, Inc., 1970.

Stauffer, Russell G. *Teaching Reading as a Thinking Process.* N.Y.: Harper & Row, 1969.

Stauffer, Russell G. *The Language-Experience Approach to the Teaching of Reading.* N.Y.: Harper & Row, 1970.

Torrance, E. Paul. *Guiding Creative Talent.* Englewood Cliffs, N.J.: Prentice-Hall, 1962.

Veatch, Jeannette. *How to Teach Reading with Children's Books.* N.Y.: Citation Press, 1968.

Veatch, Jeannette. *Individualizing Your Reading Program.* N.Y.: G. P. Putnam, 1959.

Walter, Nina W. *Let Them Write Poetry.* N.Y.: Holt, Rinehart and Winston, 1964.

Whitehead, Robert. *Children's Literature: Strategies of Teaching.* Englewood Cliffs, N.J.: Prentice-Hall, 1968.

Zirbes, Laura. *Spurs to Creative Teaching.* N.Y.: G. P. Putnam, 1959.

Index

Index